Electric Circuits

Inventive Physical Science Activities

GEMS® Teacher's Guide for Grades 3–6

by
John Erickson and **Kevin Beals**
with **Florence Stone**

Skills

Observing • Inferring • Comparing • Visualizing • Communicating • Inventing
Drawing Conclusions • Journal Writing • Using Appropriate Tools and Techniques
Making and Interpreting Circuit Diagrams • Designing a Solution or Product
Implementing a Proposed Design • Describing the Process of Technological Design
Evaluating a Product or Design

Concepts

Electricity • Electric Current • Energy Transfer • Electrical Safety
Simple Closed and Open Circuits • Series and Parallel Circuits • Short Circuits
Switches • Conductors and Insulators • Voltage • Resistance

Themes

Systems and Interactions • Energy • Structure

Mathematics Strands

Number • Logic and Language • Discrete Mathematics

Nature of Science and Mathematics

Creativity and Constraints • Science and Technology • Real-Life Applications
History of Science • Science as a Human Endeavor • Cooperative Efforts
Scientific Community • Interdisciplinary Connections

Time

Part I: Basic Circuit Activities

Four 45–60 minute sessions; one 45–90 minute session; and one 45–120 minute session

Part II: More Circuits and More *About* Circuits

Four 45–60 minute sessions; one 60–180 minute session

Great Explorations in Math and Science
Lawrence Hall of Science
University of California at Berkeley

Lawrence Hall of Science,
University of California,
Berkeley, CA 94720-5200

Director: Elizabeth K. Stage

Cover Design: Lisa Klofkorn
Internal Design and Illustrations:
Lisa Klofkorn, Stefan Gutermuth
Photography: Dan Krauss, except
as noted

Director: Jacqueline Barber
Associate Director: Kimi Hosoume
Associate Director: Lincoln Bergman
Mathematics Curriculum Specialist:
Jaine Kopp

GEMS Network Director:
Carolyn Willard
GEMS Workshop Coordinator:
Laura Tucker
Staff Development Specialists:
Lynn Barakos, Katharine Barrett, Kevin
Beals, Ellen Blinderman, Gigi Dornfest,
John Erickson, Stan Fukunaga, Linda
Lipner, Karen Ostlund
Distribution Coordinator:
Karen Milligan
Workshop Administrator: Terry Cort
Trial Test and Materials Manager:
Cheryl Webb
Financial Assistant: Vivian Kinkead
Distribution Representative:
Fred Khorshidi
Shipping Assistant: Justin Holley
Director of Marketing and Promotion:
Steven Dunphy
Principal Editor: Nicole Parizeau
Editor: Florence Stone
Principal Publications Coordinator:
Kay Fairwell
Art Director: Lisa Haderlie Baker
Senior Artists: Carol Bevilacqua,
Lisa Klofkorn

Staff Assistants: Marcelo Alba, Kamand
Keshavarz, Eyad Latif, Andrew Lee
Contributing Authors: Jacqueline
Barber, Katharine Barrett, Kevin Beals,
Lincoln Bergman, Susan Brady, Beverly
Braxton, Mary Connolly, Kevin Cuff,
Linda De Lucchi, Gigi Dornfest, Jean C.
Echols, John Erickson, David Glaser, Philip
Gonsalves, Jan M. Goodman, Alan Gould,
Catherine Halversen, Kimi Hosoume,
Susan Jagoda, Jaine Kopp, Linda Lipner,
Larry Malone, Rick MacPherson, Stephen
Pompea, Nicole Parizeau, Cary I. Sneider,
Craig Strang, Debra Sutter, Herbert Thier,
Jennifer Meux White, Carolyn Willard

Initial support for the origination and publication of the GEMS
series was provided by the A.W. Mellon Foundation and the
Carnegie Corporation of New York. Under a grant from the
National Science Foundation, GEMS Leaders Workshops were
held across the United States. GEMS has also received support
from: the Employees Community Fund of Boeing California
and the Boeing Corporation; the people at Chevron USA; the
Crail-Johnson Foundation; the Hewlett Packard Company; the
William K. Holt Foundation; Join Hands, the Health and Safety
Educational Alliance; the McConnell Foundation; the
McDonnell-Douglas Foundation and the McDonnell-Douglas
Employee's Community Fund; the Microscopy Society of
America (MSA); the NASA Office of Space Science Sun-Earth
Connection Education Forum; the Shell Oil Company
Foundation; and the University of California Office of the
President. GEMS also gratefully acknowledges the early
contribution of word-processing equipment from Apple
Computer, Inc. This support does not imply responsibility for
statements or views expressed in publications of the GEMS
program. For further information on GEMS leadership
opportunities, or to receive a publications catalog and the
GEMS Network News, please contact GEMS. We also welcome
letters to the *GEMS Network News.*

International Standard Book Number: 0-924886-74-9

Printed on recycled paper with soy-based inks.

Library of Congress Cataloging-in-Publication Data

Erickson, John, 1961-
   Electric circuits : inventive physical science activities / by John
Erickson and Kevin Beals ; with Florence Stone.
      p. cm.
   ISBN 0-924886-74-9 (trade paper)
   1. Electric circuits. I. Beals, Kevin. II. Stone, Florence, 1958-
III. Title.
QC601.E65 2004
621.3--dc22

                                                   2004002430

ACKNOWLEDGMENTS

One lucky day, Alan Gould, a long-time teacher and coordinator of physics and astronomy education, and the director of the Holt Planetarium at the Lawrence Hall of Science, noticed that an ordinary C-cell battery fits very neatly into a standard plastic film canister. Alan's discovery provided a sturdy and inexpensive (almost free) alternative to the commercial battery holder—and overcame a significant obstacle to easy and effective do-it-yourself electric projects.

Amy Waterman, a teacher at Anna Yates Elementary School in Emeryville, California, was kind enough to let us conduct the initial classroom test of this unit with her fourth-grade class. Trial-test teacher Teena Staller allowed us to use some of her photographs in the guide.

Adina Honniball, who works with science teacher Mary Chin of Berkeley Arts Magnet School in Berkeley, California, was kind enough to let us shoot photos while she taught Alex Langford's fourth-grade class. Adina also lent a hand or two and inspired the Creative Circuits "Going Further" activity on page 98.

GEMS Associate Director Lincoln Bergman helped out by conducting research on electrical inventors and contributing information on student misconceptions about electricity. Designer Stefan Gutermuth undertook the huge array of step-by-step illustrations that make the unit's instructions and examples so clear. This is the second GEMS guide photographer Dan Krauss has worked on, and we welcome his lovely work to our pages.

Stripping wires and preparing materials for the many trial-test teachers can raise a callous or two. Many thanks to Cheryl Webb, Kimi Hosoume, Vivian Kinkead, and other GEMS staff members who contributed to that effort; especially Haleah Hoshino, who also helped write the assembly instructions for some of the materials.

Many enthusiastic and supportive teachers across the country served as reviewers for *Electric Circuits* in its trial stages, and we deeply appreciate their valuable comments, classroom ideas, and samples of student work. To all the teachers and students who tested the activities in this guide, many thanks! Your insights and feedback helped us light both real and metaphoric lightbulbs and create a cohesive unit. Please see pages 250–252 for a complete list of names. ■

CONTENTS

Acknowledgments iii

Time Frame v

What You Need for the Whole Unit vi

Introduction 1

Part I: Basic Circuit Activities

Session 1: Introducing Electricity 13

Session 2: Electric Exploration Stations 35

Session 3: Simple Closed Circuits with Stations 73

Session 4: Diagrams and Short Circuits 89

Session 5: Invent a Switch 111

Session 6: Electric Switch Trade Show 121

Part II: More Circuits and More *About* Circuits

Session 7: Series Circuits 127

Session 8: Parallel Circuits 135

Session 9: Switches in Series and in Parallel 145

Session 10: Electric Inventions 161

Session 11: Electric Inventions Festival 171

Background for the Teacher 176

Acquiring and Preparing Materials 190

Teacher's Outline 227

Assessment Suggestions 239

Resources & Literature Connections 245

Reviewers 250

Depending on the age and experience of your students, the length of your class periods, and your teaching style, the time needed to present the *Electric Circuits* unit may vary. In particular, the creative nature of the two inventions sessions (5 and 10) means students will often spend a lot of time working on them if given the opportunity.

Part I: Basic Circuit Activities

Session 1: Introducing Electricity . . . . 45–60 minutes

Session 2: Electric Exploration Stations . . . . 45–60 minutes

Session 3: Simple Closed Circuits with Stations . . . . 45–60 minutes

Session 4: Diagrams and Short Circuits . . . . 45–90 minutes

Session 5: Invent a Switch . . . . 45–120 minutes

Session 6: Electric Switch Trade Show . . . . 45–60 minutes

Part II: More Circuits and More *About* Circuits

Session 7: Series Circuits . . . . 45–60 minutes

Session 8: Parallel Circuits . . . . 45–60 minutes

Session 9: Switches in Series and in Parallel . . . . 45–60 minutes

Session 10: Electric Inventions . . . . 60–180 minutes

Session 11: Electric Inventions Festival . . . . 45–60 minutes

PLEASE NOTE:
Session 11 may take more or less time than shown, depending on the expected outcome of the student projects. Time should be extended if you're inviting other classes or parents to the festival.

Some journal-writing assignments and homework time for planning and refining the student projects are not included in this time frame.

The quantities below are based on a class size of 32 students. You may, of course, require different amounts for smaller or larger classes. This list gives you a concise "shopping list" for the entire unit. Please refer to the "What You Need" and "Getting Ready" sections for each individual activity, which contain more specific information about the materials needed for the class and for each team of students.

> Before beginning the sessions in this unit, you'll first need to prepare the materials—the lightbulbs, wires, and film canisters—students will use to explore electric circuits. Be sure to read the "Acquiring and Preparing Materials" section on page 190 and refer to the "For the Homemade Electric Devices" materials listed in Consumables and General Supplies below.

Nonconsumables

For the Sessions:
- ❑ 6 two-sided copies of **Stories about Electric Accidents** (pages 50–51)
- ❑ a minimum of 32 prepared film canisters and lids
- ❑ a minimum of 36 paper fasteners (brads)
- ❑ a minimum of 30 holiday lightbulbs with stripped leads
- ❑ a minimum of 30 pieces of 4-inch-long wire with stripped ends
- ❑ 32 pieces of 4-inch-long wire with **non-stripped ends** for Station 6
- ❑ 32 pieces of 20-inch-long wire with stripped ends
- ❑ a minimum of 32 C-cell batteries
- ❑ 2 containers (anything large and strong enough) to put at the front of the room to hold finished circuit devices and batteries
- ❑ at least 5 conducting (metal) objects (such as a paper clip, bottle cap, coin, washer, bolt, metal spoon) for Station 3
- ❑ at least 5 insulating (non-metal) objects (such as a plastic spoon, straw, craft stick, piece of wood, small piece of paper, rubber band) for Station 3
- ❑ 1 container (such as a plastic bowl) to hold the stripped wires at Station 6
- ❑ 3 or more pairs of wire cutters and wire strippers
- ❑ 1 pair of pliers
- ❑ the homemade electric devices for the stations (see "Homemade Electric Devices," page 196)
- ❑ *(optional)* simple motors and buzzers

One copy each of the following posters:

- ❏ **No 120V** (page 21)
- ❏ **Battery** (page 85)
- ❏ **Lightbulb** (page 86)
- ❏ **Wire or Other Conductor** (page 87)
- ❏ **No Short Circuits!** (page 105)
- ❏ **Switch** (page 119)
- ❏ **Connection/No Connection** (page 143)

An overhead transparency of each of the following:

- ❏ **Mystery Inventor's Journal** transparencies 1–6 (pages 22–27)
- ❏ **Electrical Inventors** (page 28)
- ❏ **Lewis Howard Latimer** (page 29)
- ❏ **Light the Bulb** station sign (page 55)
- ❏ **Building Circuits from Diagrams** sheets (pages 106–107)

Copies of the following station signs:

- ❏ 1 copy of **Shocking Stories** (page 49)
- ❏ 4 copies of **Build a Circuit** (page 52)
- ❏ 2 copies of **Conductors and Insulators** (page 53)
- ❏ 2 copies of **How Many Bulbs?** (page 54)
- ❏ 2 copies of **Light the Bulb** (page 55)
- ❏ 2 copies of **Wire Stripping** (page 56)
- ❏ 1 copy each of **Quiz Board** and **Thread the Needle** (pages 57 and 58)
- ❏ 1 copy of **Repair Shop** (page 63)

Optional, but highly recommended:

- ❏ 1 copy each of **Ice Fishing, Count Up, What Matches?,** and **Upstairs/Downstairs** (pages 59–62)

Consumables

For the Sessions:

- ❏ the plug you cut off the holiday light string (see "Acquiring and Preparing Materials" on page 191)
- ❏ 1 roll of aluminum foil
- ❏ 8 sets of "Experimenter's Kits," each made up of the following materials:
  - ___ 1 strip from the **Experimenter's Kit List** (page 108)
  - ___ 1 C-cell battery
  - ___ 1 resealable plastic bag

___ 3 holiday lightbulbs with stripped leads

___ 4 pieces of 4-inch-long wire with stripped ends

___ 2 prepared film canisters and lids

___ 4 paper fasteners (brads)

❑ several scraps of cardboard, of various thicknesses

❑ various items for building (popsicle sticks, pipe cleaners, straws, etc.)

❑ 32 copies of the **Electric Invention Assignment Sheet** (page 166)

❑ *(optional)* 32 pieces of paper labeled as "Rover's Recording Sheet"

32 copies each of the following journal pages:

❑ **Electrical Inventor's Journal** (pages 30–33)

❑ **Electric Exploration Stations** (pages 64–71)

❑ **Building Circuits from Diagrams** (pages 106–107)

❑ **Switch Invention Spec Sheet** (pages 117–118)

❑ **Series Circuits** (page 133)

❑ **Parallel Circuits** (page 142)

❑ **Switches in Series and Parallel Circuits** (page 158)

❑ **Invention Spec Sheet** (pages 167–168)

For the Homemade Electric Devices:

❑ 1 copy of the **Quiz Board** template (page 220)

❑ 1 copy of the **Thread the Needle** template (page 221)

❑ 1 copy of the **Ice Fishing** template (page 222)

❑ 1 copy of the **Count Up** template (page 223)

❑ 1 copy of the **What Matches?** template (page 224)

❑ 1 copy of the **Back of What Matches?** template (page 225)

❑ 1 copy of the **Upstairs/Downstairs** template (page 226)

❑ 6 pieces of stiff cardboard, 8 $\frac{1}{2}$ x 11 inches or slightly larger

❑ 8 film canisters and lids with slits cut in tops and bottoms

❑ 39 paper fasteners (brads)

❑ 8 C-cell batteries

❑ 9 pieces of 6-inch-long wire with stripped ends

❑ 1 piece of 8-inch-long wire with stripped ends

❑ 3 pieces of 9-inch-long wire with stripped ends

❑ 2 pieces of 10-inch-long wire with stripped ends

❑ 9 holiday lightbulbs with stripped leads

❑ 1 miniature aluminum pastry dish, 4–5 inches across

❑ several pieces of aluminum foil

General Supplies

For the Sessions:

- ❑ 1 red marker
- ❑ 2 pens for Station 5
- ❑ clear tape
- ❑ masking tape
- ❑ scissors
- ❑ an assortment of metal paper clips, large and small
- ❑ stickers for students to label unfinished switches or inventions
- ❑ an overhead projector
- ❑ *(optional)* a three-hole punch
- ❑ *(optional)* 1 three-ring binder or folder in which to keep journal pages
- ❑ *(optional)* several sheets of chart paper or sentence strips for writing out key concepts
- ❑ *(optional)* 32 different-colored pens for drawing circuits in journals

For the Homemade Electric Devices:

- ❑ 8 small paper clips
- ❑ 8 large paper clips
- ❑ glue or glue stick
- ❑ tape
- ❑ paring knife or scissors with sharp point
- ❑ a hole punch
- ❑ *(optional)* 1 pair of pliers
- ❑ *(optional)* clear contact paper

Materials and Preparation

Acquiring enough materials to teach a hands-on electricity unit could be prohibitively expensive. We've skirted this issue with some innovations in material. For battery holders, we use standard plastic film canisters with paper fasteners ("brads") stuck in at either end. Ordinary C-cell batteries fit very neatly into these. For lightbulbs, we cut up a string of small, white holiday lights. Not only does this provide a cheap source of many bulbs for a class, but the bulbs are also conveniently already attached to wires, which makes them much easier for students to manipulate. Take extra lengths of wire cut from the same holiday light string, throw in some cardboard and aluminum foil, and you've got most of the materials needed for the unit.

The most challenging aspect of this guide is the *preparation* of the materials the first time you teach it, although most teachers agree it's well worth the effort. The bulk of this takes place at the beginning of the unit, so plan on an initial materials "push." The second wave of preparation comes in assembling the "homemade electric devices" used primarily in Sessions 2 and 3 (see page 6 of this introduction). The good news is that when you present the unit again in the future, a good deal of your preparation will already have been done.

As always, "What You Need for the Whole Unit" (page vi) consolidates the materials you'll need for easy "shopping," and the "Getting Ready" section of each session provides clear instructions. **To make preparation of the unit's materials as straightforward as possible, we've included a special section called "Acquiring and Preparing Materials" (page 190).** This section provides specifics on acquiring and handling the materials, and several tips for streamlining. A subsection (page 196) specifically addresses preparation of the homemade electric devices, with step-by-step directions and suggestions for ways to reduce the teacher's preparation time. As in many other activity-based science units, the preparation load can be lessened considerably if you collaborate with one or more other teachers. Students, parents, grandparents, or other adult volunteers can also provide much assistance.

Why Teach about Electric Circuits?

As part of a project by the Annenberg Foundation, new graduates from M.I.T. (the Massachusetts Institute of Technology) were informally surveyed to see whether they could light a lightbulb using a wire and a battery. Although they all thought they could do it, many of those surveyed—all graduates from one of the world's top science and engineering universities—had trouble.

Although electric appliances are a part of everyday modern life, many people have little idea, or wrong ideas, about how electricity works. (See the "Misconceptions about Electricity" section in "Background for the Teacher," page 180.) The activities in this unit provide repeated experiences and open-ended opportunities to explore basic and more advanced concepts in electric circuits.

Electric Circuits
and
National Content Standards

This guide focuses on electric circuits, including simple closed circuits; circuits in series and in parallel; and insulators and conductors. The unit supports the following content objectives in the *National Science Education Standards:* Science as Inquiry, Physical Science, Science and Technology, and History and Nature of Science. As its subtitle suggests, *Electric Circuits* holds as a central theme the concept of invention, which robustly supports *NSES* content standards. Incorporation of past and contemporary inventors, their work, and their journals (Thomas Edison's in particular) makes strong connections to the history-of-science standards and creates awareness of multicultural contributions to science over time. In particular, *Electric Circuits* directly addresses the following concepts:

- Electric circuits require a complete loop through which an electric current can pass.

- Electricity in circuits can produce light, heat, and other effects.

- In problem identification, students should develop the ability to explain a problem in their own words and identify a specific task and solution related to the problem.

- Students should make proposals to build something or get something to work better; they should be able to describe and communicate their ideas.

- Students should develop abilities to work individually and collaboratively, and to use suitable tools, techniques, and quantitative measurements when appropriate. Students should demonstrate the ability to balance simple constraints in problem solving.

- Students should evaluate their own results or solutions to problems, as well as those of other students, by considering how well a product or design met the challenge to solve a problem.

- Men and women have made a variety of contributions throughout the history of science.

In addition, this guide directly addresses the following physical-science concepts represented in one or more state science content standards:

- Electricity and magnetism are related effects that have many useful applications in everyday life. As a basis for understanding this concept:

  a. Students know how to design and build simple series and parallel circuits by using components such as wires, batteries, and bulbs.

  b. Students know electrical energy can be converted to heat, light, and motion.

- Students can demonstrate that electricity can flow in a circuit and can produce heat and light....

- Students can trace how electrical current travels by creating a simple electric circuit that will light a bulb.

Many state standards, like the national standards, include strands relating to experimentation and investigation; technology and invention; and the history of science. These are all represented in this GEMS unit.

Teaching This Unit to Older and Younger Students

The activities in *Electric Circuits* are divided into two main parts. Part I is for all students within the grade range (3–6), and covers the most basic ideas of an electric circuit. Part II is for classes ready to explore more advanced topics and create circuits with more open-ended possibilities. The entire unit works well taught at one time, but it can also be divided up, with Part I taught in one grade level, then Part II (with a review or repetition of Part I) taught at a subsequent grade. (See also "'People Model' Activities for More Advanced Students," on page 5.)

Part I: Basic Circuit Activities (grades 3–6)

The first six sessions focus on the concepts of **conductors, insulators, simple closed circuits,** and **inventing switches.** These sessions can be used by themselves as a unit suitable for third or fourth grade.

Advanced groups may show a solid grasp of the main concepts after just the first session—in which case the subsequent five sessions may simply be an opportunity for application, reinforcement, and assessment. More likely, there will be many students in a class who won't grasp the main ideas until they've participated fully in these explorations.

Part II: More Circuits and More *About* Circuits (grades 5–6)

The last five sessions concentrate on the concepts of **series circuits, parallel circuits,** and **creating electrical inventions.** These sessions are suitable for fifth- and sixth-graders who've completed the first six sessions of the unit.

Key Concepts

The key concepts addressed throughout the unit are listed in the individual overviews of the sessions in which they're introduced. Some sessions have many concepts listed; others, more focused on exploration and/or application, have none listed. They're included to help clarify the goals of each lesson for the teacher, but may also be incorporated into the lessons to reinforce the concepts for the students. You may choose to write some or all of them out in large, bold letters on separate sheets of chart paper or sentence strips, and post them on the classroom wall as they are each introduced.

Electrical Inventor's Journal

Throughout the unit each student keeps a journal, much as professional inventors do. This important concept is introduced by looking at a few examples from the pages of what students will discover to be Thomas Edison's journal.

For the students, the journals serve as a place for:

- organized storage of their data sheets, drawings, and descriptions of their inventions

One teacher assigned a key concept to each team of four students, then challenged them to present it to the class in an interesting way. To encourage the class to pay attention, she gave an award to anyone who could reproduce another's poster. Another teacher had her students copy the key concepts down in their journals.

Reading and critiquing real (or fictional, but realistic) scientists' journals can be highly motivating to students and help develop their own scientific thinking and literacy abilities. These "second-hand investigations," especially when combined with "first-hand investigations" in the classroom, show educational promise.

- recording questions and discoveries

- explaining their thinking

- getting feedback on their planning

- drawing conclusions and analyzing

- developing written communication skills and using new vocabulary

In addition, gaining familiarity with language, conventions, voice, and vocabulary that parallel actual scientists' journals can have important benefits to students—in both science and literacy development.

For the teacher, the journals permit you to:

- assess students' prior knowledge

- assess whether they're grasping the concepts

- discover what questions they have

- see what they're planning and how projects are progressing

- provide feedback on projects in progress

"People Model" Activities for More Advanced Students

It's not the primary aim of this guide to teach the structure of matter or the atomic and subatomic models of electric phenomena. Attempting to teach those concepts with very young students can lead to further misconceptions. Furthermore, teaching more advanced concepts sometimes occurs at the expense of the practical lessons of electric circuitry that students can apply directly in the classroom and in their everyday experiences.

For more advanced students, however, we've included activities called "people models," in which students act out what's happening at a subatomic (electron) level. When assembling electric circuits, we get to see *evidence* of electric current—but of course can never *see* the current itself, so it remains an abstract idea. Some students have a difficult time understanding concepts of current flowing through different kinds of circuits. Other students may be quite agile at assembling working circuits of different types, but have a hard time explaining how they work. The optional people models at the ends of Sessions 4 and 9 help make these abstract concepts more concrete. (Even the models, however, don't get into what atoms *are;* the activities are intended primarily to reinforce key concepts regarding circuits.)

The first people model addresses concepts from Sessions 1–4, and the second, concepts from Sessions 5–9. The modeling activities could also be broken down into shorter activities and done at the end of each session, dealing only with the concepts from that particular session.

Homemade Electric Devices

Six homemade electric devices are used at learning stations in Sessions 2 and 3, in addition to the six other exploration stations. Two of the devices, Quiz Board and Thread the Needle, are necessary for presenting this unit. The other four devices and learning stations—Ice Fishing, Count Up, What Matches?, and Upstairs/Downstairs—are optional, but very **strongly** recommended.

The homemade electric devices are simple toys and games that give students a mental or physical challenge, and provide important opportunities for students to explore pre-built devices and how each is made and how it works. *Ideally, all the devices should be prepared before and incorporated into Sessions 2 and 3,* as well as available for students to refer to throughout the unit.

These devices are intended to give students ideas for the projects they'll undertake in Sessions 5 and 10 (Invent a Switch and Electric Inventions). For that reason, the devices use only the kinds of materials that will be available to students during those sessions. They shouldn't be too polished looking; ideally, a student who uses one of the devices will decide that she could make a better one herself. When you teach this unit again, you can, of course, reuse these same devices. You might also consider using devices your students invent in Session 10 (if you can persuade them to part with them).

See the special section within "Acquiring and Preparing Materials" (page 196) for instructions on preparing the homemade devices you'll need for the unit, and tips for their preparation.

Summary of Sessions

Part I: Basic Circuit Activities

Session 1: Introducing Electricity

The students are introduced to the idea of an inventor's journal, with examples from Thomas Edison's. This glimpse of scientific history inspires them to use their own **Electrical Inventor's Journals** throughout the unit. An important safety message about not using 120V electricity in their experiments is introduced. They're then challenged to light a bulb using a battery and a holiday light with wires attached, to demonstrate to each other how they did it, and to draw a picture of their device in their journals.

Session 2: Electric Exploration Stations

The students begin exploring a variety of electric research stations, which provide them opportunities to become familiar with the concepts that will be introduced throughout the unit. They investigate simple and series circuits, learn about conductors and insulators, and read stories about electric safety. Four optional but *strongly* recommended "homemade electric device" stations allow students to explore devices made from the same simple materials available to them in the classroom. (See the description of these stations on page 38.)

Session 3: Simple Closed Circuits with Stations

After their explorations in Session 2, the students are ready for some concept introduction. After a discussion about the roles of the battery, conductors, and insulators in the electric devices, the entire class is challenged to work together to make a lightbulb light up. Every student is involved, by either holding a piece of wire, a lightbulb, or a battery. Once they're successful, the concepts of simple circuit, open and closed circuits, and increased voltage are introduced. More electric safety is discussed then students are introduced to the symbols used in schematic diagrams. Students then rotate through the same stations as before, but with a different perspective: to apply these newly acquired concepts to the stations as they conduct "circuit hunts."

Session 4: Diagrams and Short Circuits

The teacher leads further discussion of electric circuits, debriefing two of the stations. Each of the homemade electric devices is debriefed, providing more opportunity to discuss examples of circuits and providing inspiration for future electric inventions (if you've chosen to do Sessions 10 and 11). Students are introduced to the concept of a short circuit and—to learn more about circuits and schematic diagrams—build circuits from diagrams.

Session 5: Invent a Switch

The class is introduced to the idea of switches and how they're represented in a schematic diagram. Students are then challenged to "invent" their own switch design and add it to their circuit devices. Spec sheets are introduced and explained, and students are told they'll fill out one for their switch inventions.

Session 6: Electric Switch Trade Show

In this simulated trade show, students show off their switch inventions and compare how different teams used various conducting and insulating materials to turn their lights on and off. Students alternate in the roles of "Rover" and "Explainer," teaching other students about their own switch and learning about theirs. If you decide your class would have difficulty grasping the material in the second half of the guide, you'll end the unit here.

Part II: More Circuits and More *About* Circuits

Session 7: Series Circuits

Students begin by looking at schematic diagrams of two circuits. One is a simple closed circuit. The other is a *series circuit*—a circuit in which the electric current flows through more than one lightbulb. They predict which one will have brighter lights. Working in teams, students then build and test the circuits, and discover that when they add more lightbulbs to a circuit in series, the current decreases and the bulbs glow more dimly. This leads to an understanding that the brighter the light, the more current, and that a dimmer light means less current. Students are then introduced to the concept of *resistance;* that electric current is slowed and gives off energy when it flows through certain materials.

Session 8: Parallel Circuits

The class is introduced to the concept of *parallel circuits,* in which electric current is divided into different pathways and each pathway flows through only one resistant material (such as a lightbulb). Students compare a circuit with lightbulbs in parallel to a circuit with bulbs in series, first by looking at schematic diagrams, then by building the two circuits. They see that having the bulbs in parallel slows the current down less than in a series circuit, and allows the bulbs to burn more brightly. They also learn that when they remove one lightbulb from the series circuit, the other bulb goes out, while when they remove one from the parallel circuit, the other continues to shine. Students are given the chance to design more complex circuits and practice making schematic diagrams.

Session 9: Switches in Series and in Parallel
In this session, students apply their ability to build circuits and understand schematic diagrams. To test their predictions of the behavior of circuits with lightbulbs and switches that have connections in series and in parallel, student teams build each kind of circuit from schematic diagrams. There's a class discussion of how they tackled challenges. A valuable optional role-play of particles in a circuit (the people model) reviews all concepts introduced in Sessions 7 and 8.

Session 10: Electric Inventions
At last, students are set loose to create their own projects. In teams of two, they build a circuit or a device with a circuit in it, and write a description of what the device does and how it works, including a "spec sheet" with a schematic diagram.

Session 11: Electric Inventions Festival
This session is the highlight of the whole unit, as student pairs finish, describe, and demonstrate their projects for their classmates. (Inviting parents or other classes to the festival allows students to really show off their creations and their understanding of concepts of electric currents.) For the festival, students again become "Explainers" and "Rovers," as in Session 6. The Explainers answer questions about their devices, demonstrate how they work, and make on-the-spot repairs. The Rovers try to find out as much as they can about their classmates' projects—especially how the circuits work. ■

Part I

Basic Circuit Activities

SESSION 1: INTRODUCING ELECTRICITY

All you need to be an inventor is a good imagination and a pile of junk.... —Thomas Edison

Overview

This session, like the next, is designed to give students experience "messing around" with electric circuits. The underlying concepts won't be formally introduced until Session 3. These activities serve a variety of purposes:

- Students will begin to notice certain phenomena and patterns, and to develop intuitive ideas about electric circuits.

- They get a chance to communicate and test ideas about conductors, insulators, and electric circuits; things they've heard before or concepts they're just beginning to develop.

- They experience the "thrill of discovery."

- They become curious about how electric devices work, preparing them for the introduction of these concepts in subsequent sessions.

- They become familiar with the simple materials that will form the basis of their tinkering in all the sessions in the unit.

- They gain practice in following directions, and as independent learners.

The session begins with a look at pictures and names of inventors of electric devices. Students are informed that during this unit they will become inventors. They then look at a few pages from a famous electrical inventor's journal. From looking at the excerpted notes, students are challenged to figure out who the inventor was (Edison), and what the inventor was trying to invent (the lightbulb).

In addition to the historical connection, Edison's journal is an inspiring example to motivate students to make use of their own "inventors' journals" throughout the unit. It also introduces electrical diagrams, preparing for more in-depth diagramming activities in other sessions.

After an important safety message about **not** using 120-volt electrical outlets in their explorations, students are presented with a challenge. Given a holiday lightbulb with two stripped wire leads and a C-cell battery, students are asked to figure out how to get the bulb to light, draw a picture of their device in their journal, and to share how they

Some teachers like to post the inventor's quotes that open many sessions in this guide. They can inspire their young inventors, particularly those who are easily frustrated, or lacking in "stick-to-it-ive-ness."

Some teachers link science activities to lessons about achievements of scientists from minority groups—during Black History Month in February, for instance, or International Women's Day (March 8), or at other times of the year. See page 183 for brief summaries of the work of several such pioneers—including Lewis Howard Latimer, an African American inventor who was a colleague of Thomas Edison's. This information can be incorporated into this session and others that follow.

made the bulb light up. Through this activity they not only get to experience the fun of solving a challenge, but also begin to see what it takes to make an electric circuit through trial and error.

Before beginning the sessions in this unit, you'll first need to prepare the materials—the lightbulbs, wires, and film canisters students will use to explore electric circuits. Be sure to read the "Acquiring and Preparing Materials" section on page 190.

■ What You Need

For the class:
- ❏ 1 copy of the **No 120V** poster (page 21)
- ❏ 1 red marker
- ❏ 1 set of **Mystery Inventor's Journal** overhead transparencies 1–6 (pages 22–27)
- ❏ 1 overhead transparency of **Electrical Inventors** (page 28)
- ❏ 1 overhead transparency of **Lewis Howard Latimer** (page 29)
- ❏ the plug you cut off the holiday light string (see "Acquiring and Preparing Materials" on page 191)
- ❏ an overhead projector
- ❏ (optional) a three-hole punch

For each pair of students:
- ❏ 1 C-cell battery
- ❏ 1 holiday lightbulb with stripped leads

The C-cell batteries and light-bulbs from this session will also be used in several stations in Sessions 2 and 3.

For each student:
- ❏ an **Electrical Inventor's Journal** (pages 30–33)
- ❏ (optional) 1 three-ring binder or folder in which to keep their journal pages

■ Getting Ready

1. Copy the **No 120V** poster (page 21) and color in the circle and slash mark with a red marker. Post it in a prominent location where it can be referred to throughout the entire unit.

2. Set up the overhead projector and screen.

3. Make and set out the following transparencies: **Mystery Inventor's Journal** transparencies 1–6 (pages 22–27), **Electrical Inventors** (page 28), and **Lewis Howard Latimer** (page 29).

4. Make a copy of the **Electrical Inventor's Journal** (pages 30–33) for each student. If you've chosen to use binders, hole-punch the pages. If binders aren't an option, the journal pages can be kept in folders or stapled into booklet form—though they'll need to be added to in future sessions.

5. Set out one C-cell battery and one holiday lightbulb with stripped leads for each pair of students.

6. Have the plug you cut off the holiday light string ready to show to students during the safety message.

■ Edison's Journal

1. Place the **Electrical Inventors** transparency on the projector and ask the students if they recognize any of the people shown.

2. Tell them these are all inventors of electric devices—and that they, your students, are going to become electric inventors too during this unit.

3. Place **Mystery Inventor's Journal** Transparency **1** on the overhead projector, and say this is an excerpt from a famous inventor's journal.

4. As you read the text aloud, challenge them to try to figure out what the inventor was trying to invent, using the writing and drawings as clues. Don't divulge the correct response yet.

5. Agree that it's hard to figure the answer out from just this excerpt. Point out that the "no good" gives the impression that the inventor was testing something that didn't work.

6. Repeat the process with the remaining **Mystery Inventor's Journal** transparencies:

    a. Transparency **2**: Draw the class's attention to the words "non conductor" and "incandescent." Point out the two box-like devices shown in figure 11, and the lines that connect them.

    b. Transparency **3**: This page will offer them no clues, but is included to show that sometimes inventors have to do a lot of mathematical calculations to figure something out.

    c. Transparencies **4** and **5**: By this point, many of your students will probably begin to guess the invention was the lightbulb. (Lightbulbs will be going off in their heads!)

One teacher suggested setting up a "Hall of Inventors" in the room, with information about each famous inventor posted. Later in the unit, the children's faces and inventions could replace those of the famous inventors.

*This series of transparencies intentionally starts off quite vague and hard to figure out. By the time students reach transparency **5**, however, most or all of them will probably have figured out that the invention was the lightbulb.*

The pages from Edison's journal that are used as transparencies are excerpted from The Papers of Thomas A. Edison *edited by Reese V. Jenkins et al., Johns Hopkins University Press, Baltimore, 1989.*

d. Transparency **6**: Confirm that, indeed, Thomas Edison is the inventor and his research team was trying to invent a lightbulb that would work long enough to be practical. After *years* of watching many lightbulbs burn out too quickly, they took turns watching all day and night as one lasted for a long time. Read aloud the text of Edison's description of the event. This photo is of the lightbulb Edison demonstrated for the press just after his team had a breakthrough.

7. Ask why the invention of the lightbulb was such a big deal. Point out that it had a huge impact on the world, since people previously had to use candles and gas lamps.

8. Show the transparency **Lewis Howard Latimer,** and explain that he was the member of Edison's team who improved the filament to make it work even better. He also invented the threaded light-bulb socket.

9. Tell students that, as they could see from the journal excerpts, Edison and his team had to do a lot of exploring and "messing around" before they came up with a practical lightbulb. Edison's team actually worked on at least 3,000 different ideas and tested thousands of materials, including 6,000 plants and the beard hairs of his team members, before finally succeeding!

10. Tell the students that, like Edison and other inventors, they will spend a number of sessions "messing around" with wires, lightbulbs, and batteries. At the end of the unit, they'll get to design and build their own electric inventions.

■ Introducing the Electrical Inventor's Journals

1. Explain that, like other inventors, the students will be keeping scientific journals over the course of this unit. Ask what kinds of things they might record in their journals. Bring up the following, if they don't:

- ideas for inventions

- drawings of devices

- questions they wonder about

- descriptions of experiments, tests, and results

- descriptions of problems and solutions

- interesting stories of how they figured something out

Edison, "The Wizard of Menlo Park," (in New Jersey) patented 1,093 inventions, including the storage battery, phonograph, microphone, telephone receiver, mimeograph, and motion-picture camera. Ask your students if each invention or its "descendants" are still in use. Inventions from Edison and other inventors have a huge impact on our lives!

2. You may want to explain the steps professional inventors take to protect their work. (Your students don't need to do this unless they want to.) Sometimes inventors have to prove that they invented something and when they invented it—especially if another person claims credit for the same invention. This is one reason professional inventors keep records of what they do, and why inventor's logs usually include the time and date. They usually write in ink, so no one can erase their notes. If they make a mistake, they cross it out and write their initials next to it. This way they have a record of everything they did, including their mistakes. These journals can also be useful in obtaining a patent for an invention.

A patent is an official record that allows no one else to take credit for your idea.

3. Let students know your expectations for their journal entries. Say that when you review their journals, you'll mostly be looking for evidence in their work that they understand the content being taught in class and are completing assigned tasks. Add any other criteria you choose. (If you already know how much the journals will figure into students' grades during the unit, let them know now.)

4. Hand out the journals, and instruct the students to put them in binders (if that's the format you've chosen). Have students write their names on the cover page.

5. Ask them to open their journals to the first page, where it says, "What I know about electricity" and "Things I wonder about electricity." Give them about 10 minutes to write everything they can think of about the topic along with any questions they have.

Students' responses at this stage let you assess what they know, what misconceptions they may have, and what they're curious about. It also lets both you and them later evaluate how much they've learned through the unit.

■ An Important Safety Message

1. Remind students that in this unit they'll be making electric devices with wires, lightbulbs, and batteries. Tell students each battery has between only one and two volts. That is a safe voltage. They can touch the battery-operated devices in the unit without fear of being shocked.

2. Let them know they cannot get electrocuted from a C-cell battery, but the heat from a short circuit could cause discomfort or even a minor burn.

3. Also let them know they'll learn about short circuits later in the unit, but for now, **if they feel a battery getting hot, they should immediately disconnect the device.**

A new C-cell battery generally has a voltage of 1.5, but as it runs down, this decreases until it reaches zero. The use of one or two batteries will not pose a safety issue as long as no other dangerous conditions are present. The primary factor is the amount and path of current (measured in amps) passing through the body, and that depends on the ratio of the voltage to the resistance of the body. Wet skin has a much lower resistance than dry skin—that's why appliances that may not have very high voltages are extremely dangerous in the shower or bath.

4. Inform students that electric devices that plug into a wall socket operate on about 120 volts. **THAT IS *FAR* TOO MUCH VOLTAGE FOR PEOPLE TO EXPERIMENT WITH SAFELY IN THE CLASSROOM!**

5. Point out the **No 120V** poster and ask students what they think it means.

6. Explain that it does *not* mean that electricity is bad, or that there is no safe way to use electricity from a plug. Point out all the useful devices in the room that plug into 120-volt sockets.

7. The poster *does* mean that the electricity from a wall socket is **never** going to be part of the electric circuits they build in this class. Emphasize that electricity from a wall socket must also **never** be used for experimenting with electricity at home. Let them know they will not be able to participate in the activities if they exhibit unsafe behaviors.

8. Show them the mangled plug you cut off the holiday light string. Ask why they think you mangled it. [To prevent anyone from using it in an outlet and getting hurt.]

9. Put the plug in your pocket and say you're going to bury it in a garbage container where it's unlikely any children will find it. Tell them if they do use holiday lights for activities at home (only with parental help), they should destroy those plugs in the same way, **after cutting them off,** and dispose of them safely.

10. Ask students what they think could happen to someone who comes in contact with 120 volts. Say they'll find out more about that at the Shocking Stories station in the next session, when they'll be doing safe activities at Electric Exploration Stations.

11. Keep the **No 120V** poster prominently displayed, and refer to it regularly during the remainder of the unit.

12. Ask students to turn to the second page of their journals. Review each part of the Electric Circuits Safety Pledge with them, then ask the students to take the pledge (read each statement, check the box, and sign at the bottom). Let them know they'll need to take the pledge in order to participate in the activities of this unit.

Circuits Small and Simple

1. Organize students into working pairs.

2. Explain that for this first activity each pair will be given a battery and a holiday lightbulb with wires on it. Their challenge is to figure out how to light the bulb, **using only these materials.**

3. Explain that what they're about to make are called *electric circuits.* Define a *circuit* simply as the path an electric current follows, and say that they'll learn more about circuits shortly.

4. Tell students that once they've figured out a way to do this as a pair, each student should draw a picture of their team's device in her journal, under "Circuits Small and Simple." The drawing should be detailed enough that someone could make the light shine just by looking at the drawing.

5. Pass out a holiday lightbulb and battery to each student pair. Let pairs discover and show each other that the bulb lights when they touch each end of the battery with one of the lead wires from the lightbulb.

6. Circulate among the groups to see that each student is being given a chance to use the materials.

7. As teams complete the task and are in need of further challenges, encourage them to **combine their equipment with another team** and light up more than one bulb. Ask if they can figure out ways to make the bulbs dimmer or brighter. Again, remind them to use only the provided materials.

8. When time is up, refocus the students and remind them that what they've just made are called electric circuits. Review that a circuit is the path an electric current follows. Check for basic understanding.

9. Collect the batteries and lightbulbs and call on students to describe how they made the bulbs light up. Have them demonstrate with what you've collected, if they wish.

10. Make sure students' names are on their journals, and collect them.

If you have English language learners in your class, or if you think your students would benefit from it, you may want to go over vocabulary (battery, lightbulb, wires, paper clips, aluminum, etc.) at the end of the session. (See the glossary in the "Background for the Teacher.") The words used throughout the unit can be placed on a "Word Wall" for the students to refer to.

Consider dimming the lights in the room, such as by turning off the lights and leaving window shades open a crack. Students enjoy the dramatic effect of tiny lights in the semi-darkness. Turning the lights on can work as a signal that it's time to move on to the next part of the activity.

■ Going Further

1. More Journal Writing

Ask your students if they have further questions or if they've figured out the answers to any of their previous questions. Give them time to write and draw in their journals about more things they wonder about electricity and what they've learned about it. Encourage them to write and draw about problems and solutions, their discoveries, ideas for future projects, and unanswered questions they still have.

2. Research Famous Electrical Inventors

Assign your students to research the inventions and stories of other inventors of electric devices through history. (See "Background for the Teacher" for thumbnail sketches of the inventors mentioned in this session.)

3. Electric Devices Brainstorm

Ask your students to brainstorm devices that use electricity.

4. A Day without Electricity

Ask your students to write an imaginary story, either about a day in the life of a person before controlled electricity was invented, or a modern-day story of what life might be like if electricity was suddenly unavailable.

No 120V

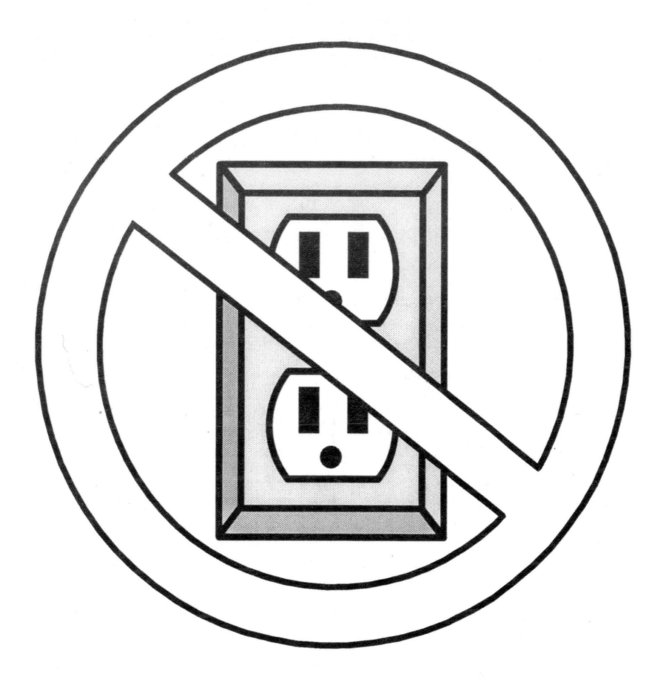

Mystery Inventor's Journal
Transparency 1

[Menlo Park,] Sept 30 1878

When[a] held in litght gets red but not quite so much as plati-na   it however keeps its redness 3 times as long as the platina. When held in powder[b] small spiral heated by 4 cells **carbon no good**[c]

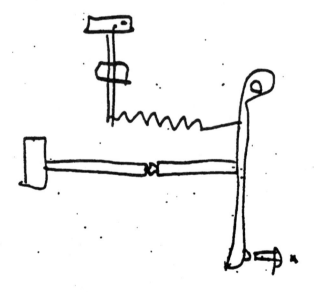

Mystery Inventor's Journal
Transparency 2

Menlo Park Oct 3 1878

fig 10

Fig 10 shews a spiral filled with some non conductor which becomes incandescent at a low heat

fig 11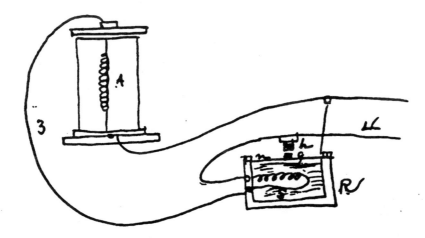

Mystery Inventor's Journal
Transparency 3

[Menlo Park,] Nov 6. 1878

try it
$3'$ wire = 7.0686 area

$$\frac{6}{42.4116^b}$$

7.35 = 42.42
8" wi = 50.2656

41.85—7.3
42.42   7.35
43.00   7.4

6 of 3 = w

$2 = \sqrt{3.^2 + 3^2} = \sqrt{9 + 9} = \sqrt{18} = 4.24$

$4 = \sqrt{4.24^2 + 4.24^2} = \sqrt{17.97 + 17.97} = \sqrt{35.9436} = 6.$

$6 = \sqrt{6.^2 + 4.24^2} = \sqrt{36 + 178} = \sqrt{54} = 7.34$

$$\begin{array}{cc} .7854 & .7854 \\ .6 & 4 \\ \hline 4.7124 & .31416 \end{array}$$

4.712   2.40

$\dfrac{1}{8} = \dfrac{9}{16} =$ .012272

$$\frac{6}{.073632}$$

.125   .125
8
1000

6 spirals .0053 = 1 spiral of .008

$2 = \sqrt{3.25^2 + 3.25^2} = \sqrt{10.56 + 10.56} = \sqrt{21.12} = 4.60$

$4 = \sqrt{4.60^2 + 4.60^2} = \sqrt{21.16 + 21.16} = \sqrt{42.32} = 6.50$

$6 = \sqrt{6.50^2 + 4.60^2} = \sqrt{41.25 + 21.16} = \sqrt{62.41} = 7.90$

m.

nearly $\dfrac{8}{} =$

Mystery Inventor's Journal
Transparency 4

Menlo Park Oct 3 1878

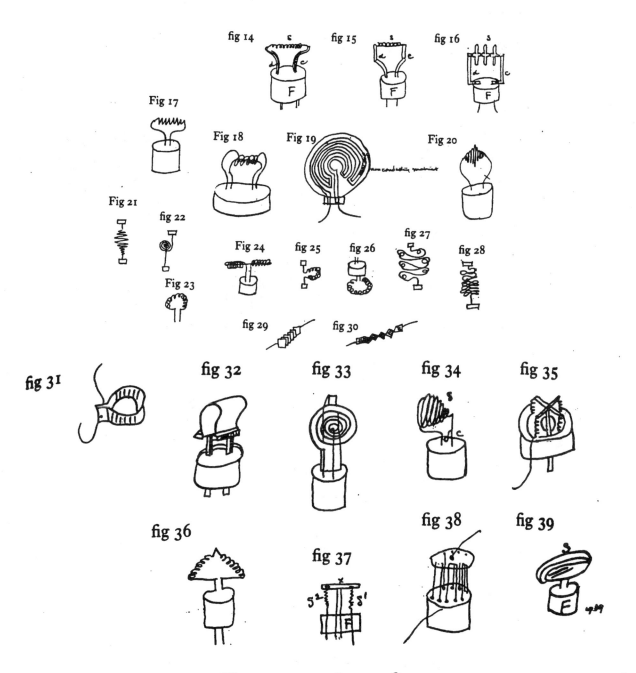

Figs. 14. 15. 16. 17. 18. 19. 20. 21. 22. 23. 24. 25. 26. 27. 28. 29. 30. 31 & 32 33. 34. 35. 36. 37. 38. & 39 are different forms in which I propose to experiment with in ascertaining the best form for giving light by incandescence[b] due to the passage of the Electric current through them

Mystery Inventor's Journal
Transparency 5

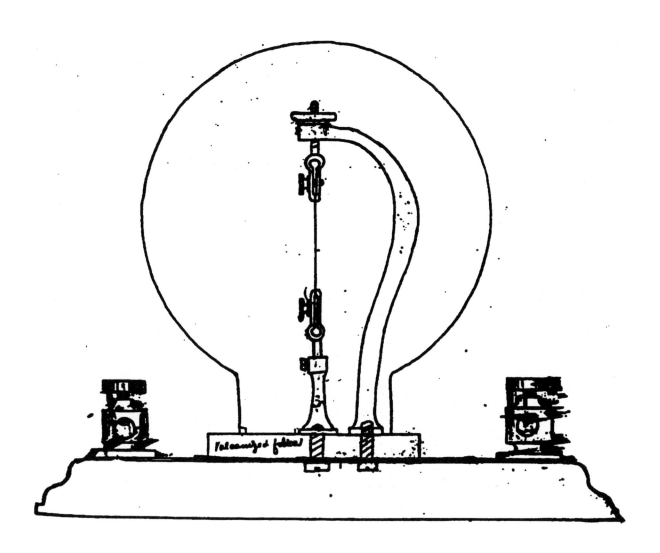

Mystery Inventor's Journal
Transparency 6

'We sat and watched it with anxiety growing into elation. It lasted about forty-five hours, and then I said, "If it will burn that number of hours now, I know I can make it burn a hundred!'

October 21, 1879, has ever since been celebrated the world over as Edison Lamp Day.

Electrical Inventors

Lewis Howard
Latimer

Grace Murray
Hopper

Thomas Edison

Nikola Tesla

Hertha Marks
Ayrton

Granville Woods

Michael Faraday

Otis Boykin

Mark Dean

Alexander
Graham Bell

Samuel Morse

Lewis Howard Latimer
(1848–1928)

Lewis Latimer, a member of Edison's team, improved the lightbulb filament. He also invented the threaded ("screw-in") lightbulb socket.

Electrical Inventor's Journal

ideas for inventions • drawings • questions • experiments • problems • solutions • stories

Inventor's Name: _____

Electrical Inventor's Journal

ideas for inventions • drawings • questions • experiments • problems • solutions • stories

What I know about electricity.

_____

_____

_____

_____

_____

_____

_____

_____

_____

Things I wonder about electricity.

_____

_____

_____

_____

_____

_____

_____

_____

_____

_____

1

Electric Circuits Safety Pledge

Read each of the statements below, make a check in the box to show you've read it, then sign your name at the bottom.

- I understand that power lines and electric outlets can be very dangerous to people and could lead to death or serious injury. ☐

- I promise never to experiment with electric outlets, and never to experiment with appliances that are plugged in or will be plugged in again. ☐

- I promise never to put anything inside electric outlets that wasn't designed to be there. ☐

- I promise not to touch power lines. ☐

My signature: _____

(2)

ideas for inventions • drawings • questions • experiments • problems • solutions • stories

Circuits Small and Simple

Get the bulb to light, then draw your circuit device.

③

In reference to the 10,000 times he didn't succeed in his storage battery experiments, Thomas Edison said, *Why, I have not failed. I've just found 10,000 ways that won't work.*

Overview

In this session, the students are first introduced to the Electric Exploration Stations and how they'll circulate among them and record results. There are eight main stations and four optional, but highly recommended, stations.

Two aspects of this activity are worth noting.

1. Flexible station rotation format.

The Electric Exploration Stations in Sessions 2 and 3 are designed to be used in a flexible format, with student pairs moving to new stations when they've completed the activities at others, rather than on a timed rotation. This allows for:

- each station activity's time requirements (With a timed rotation, some students end up being rushed from a station before finishing, and others are forced to remain at a station after they've finished the activity.)

- differing student ability levels and working pace

- opportunity for students to pursue their own interests

2. It's expected that the students will not complete all the stations in Session 2.

The stations in Sessions 2 and 3 are the same, but the approach in each session is slightly different. In Session 2, each pair of students will likely complete three to five stations. This time at the stations serves largely as an opportunity to explore electricity, an important phase of the learning cycle meant to inspire questions and ideas.

We recommend that you read ahead to Sessions 10 and 11 (pages 161 and 171) to decide whether you'll do these with your class. This will affect your instructions to the students about the device stations.

Preliminary Key Concepts for Session 2

Note: These concepts will be taught more formally in Session 3, but students will begin to encounter them during this session.

- Energy can be carried from one place to another by electric current.

- Electric circuits require a complete loop through which an electric current can pass.

- Electricity in circuits can produce light and heat.

- Energy can be stored in batteries.

- Electric current flows through some materials (conductors) and does not flow through other materials (insulators). Metals are electric conductors.

- When an electric circuit is open, the particles stop flowing and the current stops; when an electric circuit is closed, the current flows.

■ What You Need

For the class:
- ❏ the **No 120V** poster from the previous session

For Station 1: Shocking Stories
- ❏ 1 copy of the **Shocking Stories** station sign (page 49)
- ❏ 6 two-sided copies of **Stories about Electric Accidents** (pages 50–51)

For Station 2: Build a Circuit
- ❏ 4 copies of the **Build a Circuit** station sign (page 52)
- ❏ 16 prepared film canisters and lids (see "Acquiring and Preparing Materials" on page 195)
- ❏ 32 paper fasteners (brads)
- ❏ 16 holiday lightbulbs with stripped leads
- ❏ 16 pieces of 4-inch-long wire with stripped ends
- ❏ 16 C-cell batteries
- ❏ 2 containers (anything large and strong enough) to put at the front of the room to hold finished circuit devices and batteries

For Station 3: Conductors and Insulators
- ❏ 2 copies of the **Conductors and Insulators** station sign (page 53)
- ❏ at least 5 conducting (metal) objects (such as a paper clip, bottle cap, coin, washer, bolt, metal spoon)
- ❏ at least 5 insulating (non-metal) objects (such as a plastic spoon, straw, craft stick, piece of wood, small piece of paper, rubber band)
- ❏ 2 prepared film canisters and lids
- ❏ 4 paper fasteners (brads)
- ❏ 2 holiday lightbulbs with stripped leads
- ❏ 2 pieces of 4-inch-long wire with stripped ends
- ❏ 2 C-cell batteries

For Station 4: How Many Bulbs?
- ❏ 2 copies of the **How Many Bulbs?** station sign (page 54)
- ❏ 10 holiday lightbulbs with stripped leads
- ❏ 2 C-cell batteries

For Station 5: Light the Bulb
- ❏ 2 copies of the **Light the Bulb** station sign (page 55)
- ❏ 2 holiday lightbulbs with stripped leads
- ❏ 2 C-cell batteries
- ❏ 2 pens

For Station 6: Wire Stripping
- ❏ 2 copies of the **Wire Stripping** station sign (page 56)
- ❏ 2 pairs of wire strippers
- ❏ 32 pieces of 4-inch-long wire with **non-stripped ends**
- ❏ 1 container (such as a plastic bowl) to hold the stripped wires
- ❏ (optional) extra pairs of wire strippers or scissors

For Stations 7 and 8: Quiz Board and Thread the Needle
- ❏ 1 copy each of the **Quiz Board** and **Thread the Needle** station signs (pages 57 and 58)
- ❏ the corresponding homemade electric devices (see "Homemade Electric Devices," page 196)

For the Repair Shop
- ❏ 1 copy of the **Repair Shop** station sign (page 63)
- ❏ 1 pair of pliers
- ❏ 1 pair of wire strippers
- ❏ 2 feet of aluminum foil
- ❏ tape
- ❏ extra C-cell batteries
- ❏ 1 pair of scissors
- ❏ any materials leftover from Session 1 (such as prepared film canisters, paper fasteners, stripped wires, and holiday lightbulbs with stripped leads)

Optional, but highly recommended:

For Stations 9, 10, 11, and 12: Ice Fishing, Count Up, What Matches?, and Upstairs/Downstairs
- ❏ 1 copy each of the **Ice Fishing, Count Up, What Matches?,** and **Upstairs/Downstairs** station signs (pages 59–62)
- ❏ the corresponding homemade electric devices (see "Homemade Electric Devices," page 196)

For each student:
- ❏ the **Electrical Inventor's Journal** from the previous session
- ❏ 1 copy of the **Electric Exploration Stations** journal pages (pages 64–71) to add to journal

■ Getting Ready

One Week Ahead

Prepare the homemade electric devices according to the instructions on pages 198–219.

Before the Day of the Activity

1. Setting Up the Electric Exploration Stations

There are eight main stations, plus a repair shop, and four highly recommended optional stations. The first eight stations will accommodate 32 students, but some will be crowded. Adding the four optional stations spreads your students out more, is a lot of fun, and provides more examples of different kinds of circuits that can be built from simple materials.

Main Stations

Station 1: Shocking Stories (accommodates six students)
Station 2: Build a Circuit (2 tables) (eight students, four at each table)
Station 3: Conductors and Insulators (four students)
Station 4: How Many Bulbs? (four students)
Station 5: Light the Bulb (four students)
Station 6: Wire Stripping (two students)
Station 7: Quiz Board (two students)
Station 8: Thread the Needle (two students)

Highly Recommended Optional Stations
(each accommodates two students)

Station 9: Ice Fishing
Station 10: Count Up
Station 11: What Matches?
Station 12: Upstairs/Downstairs

Use the illustration on the next page to set up the stations in your classroom. The illustration shows an overhead view of the tables and chairs, and the information included gives a brief description of each station and its objective. (Specific instructions about how to prepare each individual station follow, on page 40.) *Reminder:* **All students must have completed Stations 1 and 2 by the end of Session 3.**

Station 1: Shocking Stories
Teams read stories based on real accidents with electricity.

The stories are meant to get them to think about electric safety precautions.

Station 2: Build a Circuit
Each team builds a simple circuit, following the step-by-step directions.

The circuit devices they build will also be used

in Sessions 5 and 6. This helps them become familiar with the materials and procedure that will be used throughout the unit.

Station 3: Conductors and Insulators
Teams test materials to see which conduct electricity and which don't.

Station 4: How Many Bulbs?
Teams try to figure out ways to light more than one lightbulb with one battery.

This provides the opportunity to begin to discover series and parallel circuits, which will be explored in later sessions.

Station 5: Light the Bulb
Teams look at diagrams of how a battery, wire, and lightbulb can be connected. They predict, then test, ways to light the bulb.

The diagrams are based on common misconceptions about electricity; the experience provides students firsthand evidence to support or refute their own preconceptions.

Station 6: Wire Stripping
Teams use wire strippers to strip wire.

Students learn a valuable electrical procedure, while helping prepare materials to use in other sessions of the unit.
Note: You may choose to eliminate this station for young students who may have difficulty with it.

Station 7: Quiz Board
Teams play the game and try to figure out how the device works.

The homemade electric devices in Stations 7–12 are simple toys and games that give students a mental or physical challenge, and serve as inspiration for the projects they will undertake in Sessions 5 and 10.

Station 8: Thread the Needle
Teams play the game and try to figure out how the device works.

See Quiz Board description.

Repair Shop
Teams bring devices that aren't working to this location to try to fix them.

The "shop" is meant for students to work together and teach each other how to fix things. This both empowers them, and alleviates some of the repair burden from the teacher.

Highly Recommended Optional Stations
(set up as in Station 7)
Station 9: Ice Fishing
Station 10: Count Up
Station 11: What Matches?
Station 12: Upstairs/Downstairs
Teams play the game and try to figure out how the device works.

See Quiz Board description.

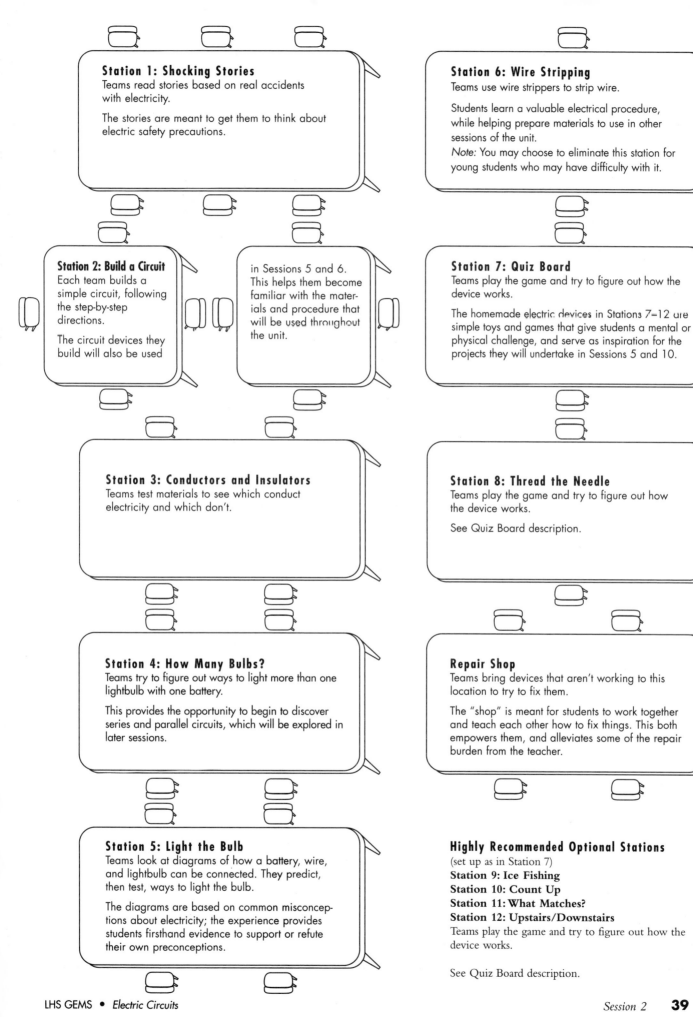

2. Preparing the Stations

For each of the following stations, designate a table (two tables for **Build a Circuit**) and set the station materials on it. Alternatively, store the materials in bags, tubs, or trays to set out when the electric exploration stations begin.

Station 1: Shocking Stories

1. Set out the copy of the **Shocking Stories** station sign (page 49).

2. Set out six copies of **Stories about Electric Accidents** (pages 50–51).

Station 2: Build a Circuit

1. Set out two copies of the **Build a Circuit** station sign (page 52) on each of the two tables.

2. For each pair of students, set out the following materials:

   - 1 film canister and lid
   - 2 paper fasteners
   - 1 holiday lightbulb with stripped leads
   - 1 piece of 4-inch-long wire
   - 1 C-cell battery

3. Put two containers at the front of the room. Label one to hold the batteries, the other to hold the circuit devices students build at the station.

Station 3: Conductors and Insulators

1. Set out two copies of the **Conductors and Insulators** station sign (page 53).

2. Set out the conducting and insulating materials you collected.

3. Following the directions on the **Build a Circuit** station sign, make two circuit devices. Set them out at the station.

One teacher suggested making a recording of each of the "Shocking Stories" for non-independent readers to listen to.

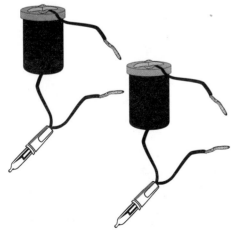

Station 4: How Many Bulbs?

1. Set out two copies of the **How Many Bulbs?** station sign (page 54).

2. Set out the following materials:

- 10 holiday lightbulbs with stripped leads
- 2 C-cell batteries

Station 5: Light the Bulb

1. Set out two copies of the **Light the Bulb** station sign (page 55).

2. Set out the following materials:

- 2 holiday lightbulbs with stripped leads
- 2 C-cell batteries
- 2 pens

Station 6: Wire Stripping

1. Set out two copies of the **Wire Stripping** station sign (page 56).

2. Set out the following materials:

- 2 pairs of wire strippers (plus any extra pairs along with some scissors)
- 32 pieces of 4-inch-long wire with **non-stripped ends**
- a container to hold the stripped wires

Station 7: Quiz Board

1. Set out the copy of the **Quiz Board** station sign (page 57).

2. Set out the Quiz Board device you prepared earlier.

Station 8: Thread the Needle

1. Set out the copy of the **Thread the Needle** station sign (page 58).

2. Set out the Thread the Needle device you prepared earlier.

Stations 9–12: Ice Fishing, Count Up, What Matches?, and Upstairs/Downstairs

1. Set out one copy each of the **Ice Fishing, Count Up, What Matches?,** and **Upstairs/Downstairs** station signs (pages 59–62).

2. At the appropriate station, set out the devices you prepared earlier.

Repair Shop

1. Set out the copy of the **Repair Shop** station sign (page 63).

2. Set out the following materials:

- a pair of pliers

- a pair of wire strippers

- aluminum foil

- tape

- C-cell batteries

- scissors

- other materials such as film canisters, paper fasteners, stripped wires, and holiday lightbulbs with stripped leads

Optional Additional Stations

If you think you need more stations to accommodate your students, here are some suggestions:

- Make duplicates of some of the stations.

- Set up a reading station with books about electricity. See "Resources" on page 245 for suggestions.

- Bring in expendable electric devices (such as flashlights), working or broken, to take apart and search for circuits. **Do not use devices that can be plugged in.**

- Set up an additional electricity safety station with safety coloring books or other materials.

- Make one or more homemade circuits with wires missing or loose, and challenge students to try to repair them.

On the Day of the Activity

1. Be sure the **No 120V** poster is still displayed prominently enough for all students to see it.

2. For each student, make a set of the eight **Electric Exploration Stations** journal pages (pages 64–71). Have them ready to distribute to students.

3. Set out the students' **Electrical Inventor's Journals** from the previous session.

■ Introducing the Electric Exploration Stations

1. Tell the students they'll be rotating through eight (or more) electric exploration stations. Point out Stations 1: Shocking Stories and 2: Build a Circuit. Tell them **these two stations are the most important,** and they'll all need to complete them in either this session or the next.

2. Point out one of the station signs and say that students will need to read each step on the signs and look at the pictures to understand how to do the activity at each station.

3. If there are any stations you think your students might find confusing—such as Station 2: Build a Circuit—briefly demonstrate difficult parts of those procedures without actually doing the activities or giving away the results. Review the concept of a circuit, if needed.

4. Explain that each student will carry her **Electrical Inventor's Journal** and a pen to record her results. Say that although they'll work in pairs to do the activities at the stations, each will write, draw, and answer questions in her own journal.

5. Point out that some stations accommodate two students, some four, and so on. Say that when they finish their first station, they should leave it as neat as (or neater than) when they found it before moving on to any other available station. Tell students to **disconnect the circuits when they leave a station,** so the bulbs and the batteries won't burn out.

6. Tell them not to rush, and that they'll have another session in which to complete the stations. Remind them that stations are set up to accommodate a certain number of students. If, for instance, there are four chairs at a station and all are occupied, they should move to another station.

■ Stations Overview

Station 1: Shocking Stories

1. Sum up the students' assignment:

- Read the stories.

- Write at least five pieces of electrical-safety advice you'd give someone who didn't know how dangerous electric current in houses and power lines can be.

2. Explain that the five "shocking stories" describe dangerous electric situations. Reassure your students that in the class they will only be conducting *safe* electric investigations, with low-voltage batteries. The stories provide examples of what they should be careful to avoid in order to be safe with electricity.

Station 2: Build a Circuit

1. Sum up the assignment:

- Follow the steps on the station sign to make a circuit.

2. Explain that, using the materials provided at the station, the students follow step-by-step directions to make a simple circuit. Let them know they'll use their circuit in other sessions as well. Because this is just a building activity, there is nothing to record in their journals, other than to check the station off.

3. Tell students that after they've built their circuit device, they'll take it to the containers at the front of the room to put the battery in one container and the assembled, but "empty," circuit in the other container.

"Build a Circuit" is the only station at which students will build something they'll use later in the unit.

Station 3: Conductors and Insulators

1. Sum up the assignment:

- Test and list at least five conductors and five insulators.

2. Let students know they will use the circuit devices at the station to test a variety of provided materials to see if they conduct electricity (conductors) or not (insulators). They may also test other materials from around the classroom. **Remind them not to use 120V outlets!**

3. Tell students they should list at least five conductors and five insulators in their journals.

Station 4: How Many Bulbs?

1. Sum up the assignment:

- Try to light up more than one bulb, using one battery.

- Draw how you did it.

2. Remind students that they already figured out how to light one bulb in Session 1 (refer them to the Circuits Small and Simple page of their journals), but now they're challenged to see if they can figure out how to light *more* than one using only one battery. If you have students who could use a greater challenge, ask them to come up with *more than one way* to light up two or more bulbs with one battery.

Station 5: Light the Bulb

1. Sum up the assignment:

- Look at the drawings and circle which ones you *predict* will make the bulb light up.

- Test them, and circle which ones *did* light up the bulb.

2. Make sure your students understand that a prediction is what someone thinks *before* something happens—and that they should *not* correct their predictions after testing. Tell them to write their predictions in ink, to help prevent this temptation. Emphasize that many college students have trouble making these predictions, and their predictions don't need to be correct.

3. Remind students that *after* they've made all their predictions, they'll use the material at the station to "build" the drawings and test whether they light up the bulb.

4. Make sure students know they are to mark their predictions and results **in their journals,** not on the station sign.

Station 6: Wire Stripping

1. Sum up the assignment:

- Strip both ends of one wire about $\frac{3}{4}$ inches.

- Put the stripped wire in the container.

2. Let students know they should squeeze the wire stripper firmly enough to cut through the plastic insulation, but not so firmly that they cut the wire beneath it. Also, let them know that each student should strip both ends of just one wire.

Stations 7 (Quiz Board) through 12 (Upstairs/Downstairs)

1. Sum up the assignment for Stations 7 and 8, and whichever of Stations 9–12 you've chosen:

- Play the game.

- Try to figure out how the device was made and how it works.

- Answer the questions.

- Get ideas.

2. Explain that there are electric games and challenges at these stations. Hold up a few so students can see they are simple homemade devices. Express thanks to anyone who helped construct them.

3. Tell students to play the games and explore each device. Encourage them to look at the front and back of each device to try to figure out how it works, but tell them they may not take anything apart.

4. For each device, students should discuss and answer these questions:

- What do you think you are supposed to do with it?

- What are the parts you can see?

- What parts do you think you **can't** see?

- How do you think it works?

5. If you'll be doing Sessions 10 and 11 of this unit, tell students they'll soon be designing their own devices, and that they might get ideas of what to make and how to put their devices together by exploring the ones they see at these stations.

Repair Shop

1. Explain that students will take anything that isn't working to this location to try to fix it.

2. Ask if anyone has ever tried to use an electric device that was broken. Explain that some of the devices at the stations could break. If students think one is broken, *they're* the repair people! If any device breaks down and can't be fixed quickly at the station, it should be taken to the "repair shop."

3. Say that not only can the students who *take* a broken device to the repair shop try to repair it, but that other students may help too. Tell them the "shop" is meant for working together and teaching each other how to fix things. Let them know that although it can be frustrating at first, they'll get better at repairing things as the unit goes on.

**Emphasize that students
SHOULD NOT
try to repair 120-volt
electric devices!**

■ Exploring the Stations

1. Remind students of the following:

   • They'll rotate through eight (or more) stations, and **must** complete Stations 1 and 2 in this or the next class session.

   • They'll travel and work in pairs, but each will write in his own journal.

   • They'll have two class sessions to complete the stations, so they needn't rush, yet they shouldn't waste time waiting to visit a station that already has enough students.

   • They should leave a station as neat or neater than the way they found it, and **disconnect circuits** when they leave.

   • They should visit the repair shop if they find something broken or need to do any sort of repairs.

2. When students understand what they are to do, organize them into teams of two. Hand out the journals from Session 1 and the eight journal pages you copied for this session. Ask students to add the **Electric Exploration Stations** pages to their journals, and point out where they should write, record, or answer questions.

3. Assign the teams to their first stations (don't assign students to the repair shop, of course) and have them begin the activities. Circulate as needed to assist. Occasionally a team may need to be "inspired" to move on, if they're unnecessarily stuck at a particular station.

4. Five or ten minutes before class time is up, announce that the students will have more time to do the stations during the next session. Tell them to spend the remainder of the session writing and drawing in their journals. Point out page 11 of their journals, and remind them to include:

- answers to some of their questions
- questions they wonder about
- ideas for inventions
- drawings of devices
- descriptions of experiment tests and results
- descriptions of problems and solutions
- interesting stories of how they figured something out

5. Collect the journals. Leave the stations up, if possible, since they'll be used during the next session.

Note: To be sure the batteries don't get accidentally drained between sessions, check that leads aren't crossed (a short circuit!) and that all circuits are firmly disconnected.

Make sure no wires are crossed or that no circuit is complete.

Shocking Stories

1. Read the Stories about Electric Accidents.

2. Write in your journal at least five pieces of advice on electrical safety you'd give someone who didn't know how dangerous the electric current in houses and power lines can be.

Stories about Electric Accidents

Shocking Story 1: Kite Flying

Benjamin Franklin did an experiment to find out if lightning was electricity. He tied a metal key to the end of a kite string, then flew the kite in a thunderstorm. Electricity traveled down the string to the key causing small sparks to jump from the key to his hand. Another scientist tried to do the same experiment, but was not so lucky. He was killed instantly when a blue ball of fire as big as a fist came out of the rod he was using and struck him in the head. Since then, many people have died or been injured when kites touch power lines or are struck by lightning.

Shocking Story 2: Electric Outlet

One day Karen was playing with her baby brother. She turned away to pick something up, and when she turned back, he was gone. When she found him, he was trying to stick a metal fork in an electric outlet, which Karen knew could kill him. As she rushed to stop him, she was relieved to see there was a plastic cover on the outlet. She took away the fork and told him "no," then told her parents about what had happened.

Shocking Story 3: Electric Appliance Near Water

Rhonda wanted to listen to music while she took a bath. She plugged in her boom box and set it on a shelf next to the tub, where she thought it would be safe. Her little sister bumped the boom box, and it fell in the tub. Rhonda was electrocuted and killed instantly.

Shocking Story 4: Short Circuit

Jamal and his four brothers and sisters lived with their aunt. As they were sleeping one night, a short circuit in their Christmas tree lights started a fire. A neighbor helped rescue his brothers and sisters, but because of smoke, couldn't get to Jamal or his aunt. When firefighters arrived, Jamal and his aunt were both unconscious. Both Jamal and his aunt later recovered, but were lucky to survive.

Shocking Story 5: Power Line

Twelve-year-old Miguel liked to explore. There was a tall electric tower near his school that he walked by every day. He wondered what it would be like to climb it. One day when no one was there, he climbed over the fence around it, and began to climb the tower. The tower was very tall, and it was exciting to climb so high. When he was 60 feet high in the tower, he touched a power line. He was killed instantly when 60,000 volts of electricity shot through his body.

1. Twist together the little wires on a holiday bulb and on a wire without a bulb.

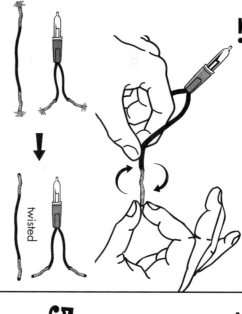

twisted

2. Wrap the stripped end of one of the wires from the bulb around the "legs" of a paper fastener just under the head.

3. Push the legs of that same paper fastener into the slit in the bottom of the film canister from the outside to the inside of the canister.

4. Reach into the canister with one finger and separate the legs of the paper fastener, squashing them flat against the bottom of the canister.

5. Push the legs of the second paper fastener through the slit in the lid from the inside to the outside of the lid.

6. Wrap one stripped end from the wire without a bulb around the legs of the paper fastener right at the top of the lid.

7. Separate the legs of the paper fastener, squashing them flat against the top of the lid.

8. Put the battery into the canister with the bump facing the bottom of the canister.

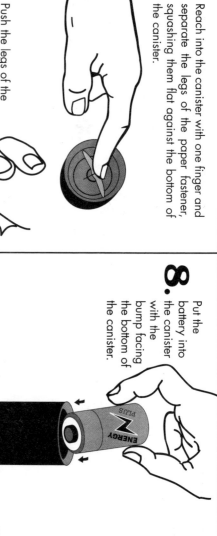

9. Put the lid on and close it tight. Touch the bulb wire with the non-bulb wire. If it doesn't light up, make sure everything is hooked up right and try again.

Station 3: Conductors and Insulators

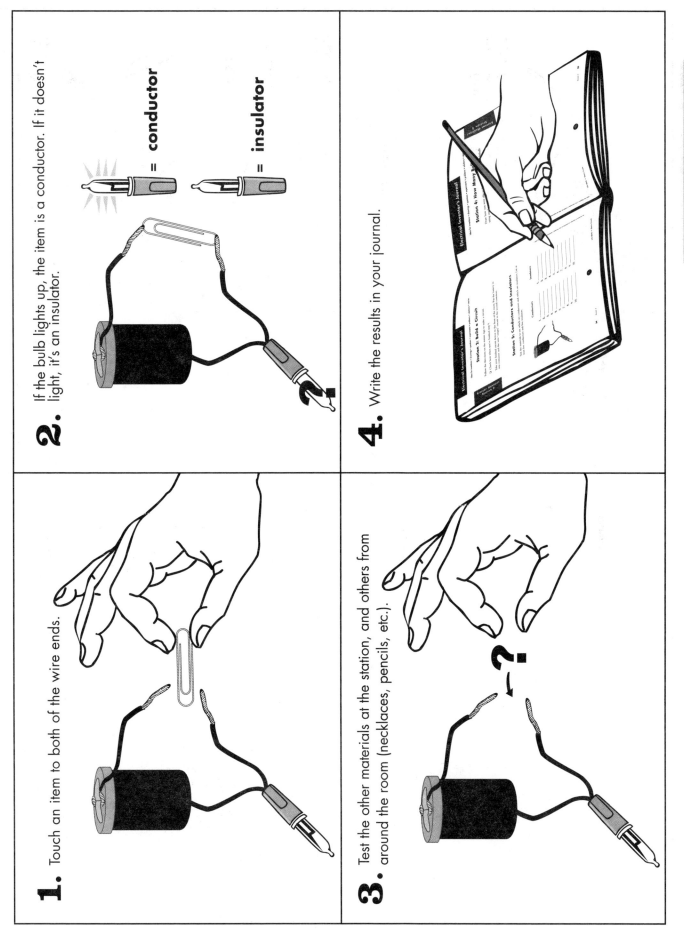

1. Touch an item to both of the wire ends.

2. If the bulb lights up, the item is a conductor. If it doesn't light, it's an insulator.

= conductor

= insulator

3. Test the other materials at the station, and others from around the room (necklaces, pencils, etc.).

4. Write the results in your journal.

Station 4: How Many Bulbs?

1. Can you light up two lightbulbs at once? Make a drawing in your journal of how you did it.

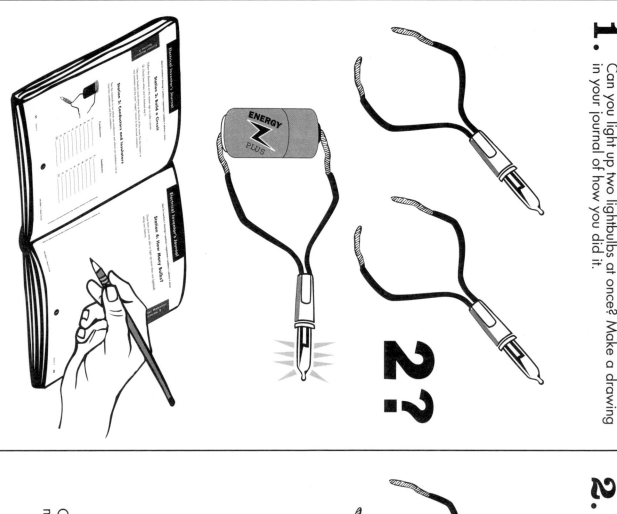

2?

2. See how many bulbs you can light at once using **one** battery.

3? 4? 5?
...

Can you figure out a different way to light up two or more bulbs with one battery?

Station 5: Light the Bulb

1. Look carefully at each drawing and think which ones would light the bulb in real life.

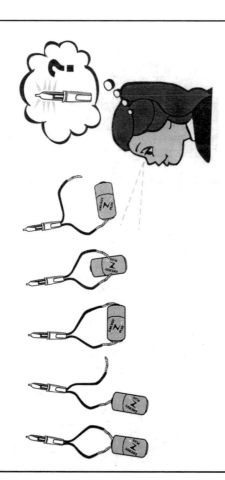

2. Circle with a pen in your journal the drawings of devices that you think will light the bulb.

I predict these will light up:

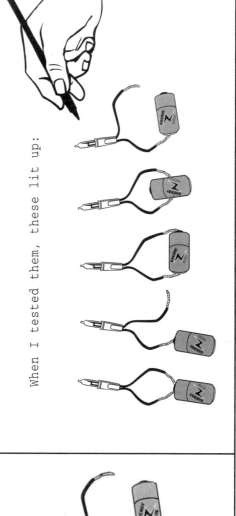

3. Use the materials to copy and try out each one of the drawings.

4. Circle any that made the bulb light up.

When I tested them, these lit up:

1. Grip the end of the wire in the wire stripper. Cut only the insulation on the outside. Do not cut the metal on the inside.

← about 3/4"

2. Pull the wire away from the stripper leaving the insulation behind.

insulation

bare metal

3. Do the same to the other end of the wire.

4. Put the stripped wire in the container.

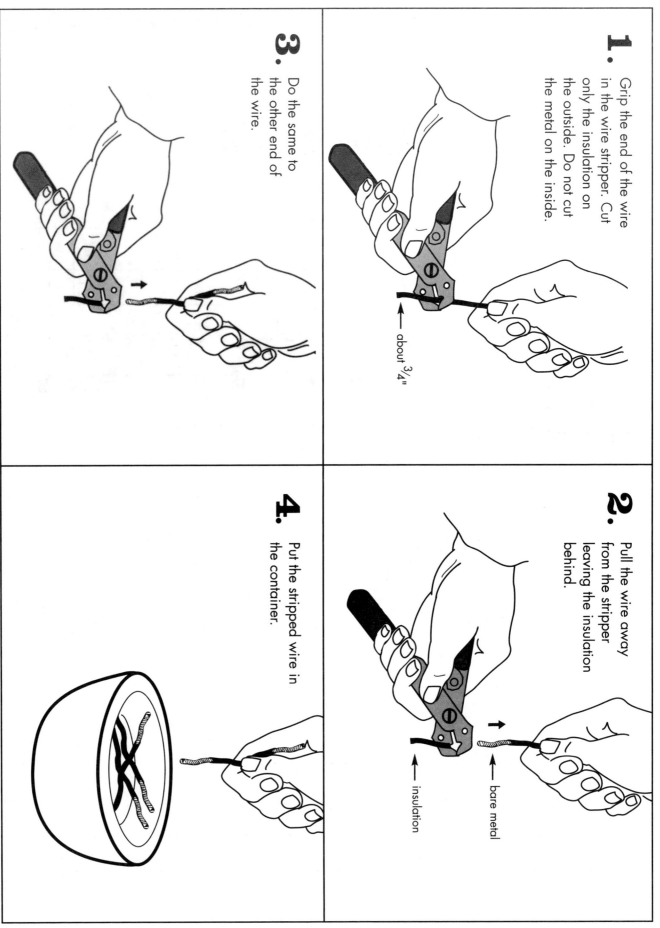

Quiz Board

1. What do you think you are supposed to do with it?

2. What are the parts you can see?

3. What parts do you think you **can't** see?

4. How do you think it works?

Thread the Needle

1. What do you think you are supposed to do with it?

2. What are the parts you can see?

3. What parts do you think you **can't** see?

4. How do you think it works?

Ice Fishing

1. What do you think you are supposed to do with it?

2. What are the parts you can see?

3. What parts do you think you **can't** see?

4. How do you think it works?

Count Up

1. What do you think you are supposed to do with it?

2. What are the parts you can see?

3. What parts do you think you **can't** see?

4. How do you think it works?

What Matches?

1. What do you think you are supposed to do with it?

2. What are the parts you can see?

3. What parts do you think you **can't** see?

4. How do you think it works?

Upstairs/Downstairs

1. What do you think you are supposed to do with it?

2. What are the parts you can see?

3. What parts do you think you **can't** see?

4. How do you think it works?

Repair Shop

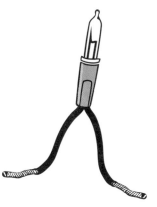

This is where you work together to try
to fix things that aren't working.

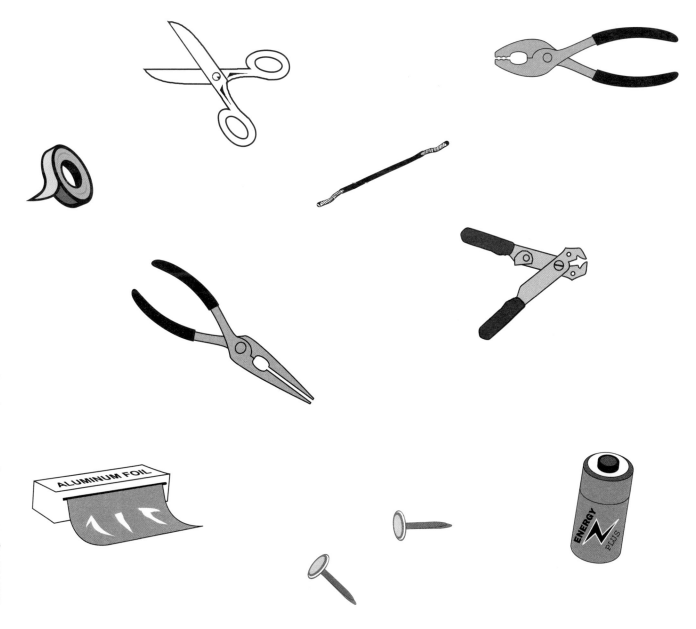

Electrical Inventor's Journal

ideas for inventions • drawings • questions • experiments • problems • solutions • stories

Electric Exploration Stations

Station 1: Shocking Stories

Note: C-cell batteries have only 1–2 volts of electricity, so they are safe to use. We will only be doing safe experiments with batteries in our classroom activities.

After reading the stories, write **at least five** pieces of advice on electrical safety you'd give someone who didn't know how dangerous the electric current in houses and power lines can be.

_____

_____

_____

_____

_____

_____

_____

_____

_____

_____

_____

_____

_____

_____

4

Electrical Inventor's Journal

ideas for inventions • drawings • questions • experiments • problems • solutions • stories

Station 2: Build a Circuit

Follow the directions on the station sign to make a circuit.

❑ Check here when you've finished step 9.

Take your finished circuit device to the front of the room. Put the battery in one container and the now "empty" circuit in the second container.

Station 3: Conductors and Insulators

Test the materials to see which are conductors and which are insulators. List at least five conductors and five insulators.

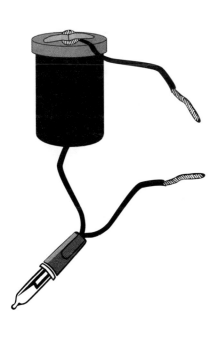

Conductors	Insulators
1. _____	1. _____
2. _____	2. _____
3. _____	3. _____
4. _____	4. _____
5. _____	5. _____
6. _____	6. _____
7. _____	7. _____
8. _____	8. _____
9. _____	9. _____
10. _____	10. _____

5

Electrical Inventor's Journal

ideas for inventions • drawings • questions • experiments • problems • solutions • stories

Station 4: How Many Bulbs?

Draw how you were able to light up more than one lightbulb using one battery.

6

Electrical Inventor's Journal

ideas for inventions • drawings • questions • experiments • problems • solutions • stories

Station 5: Light the Bulb

Using a pen, circle the ones you **predict** will make the bulb light up.

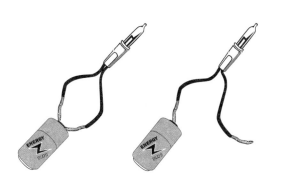

Circle the ones that actually **did** light up when you tested them.

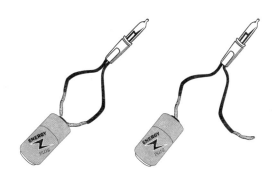

Station 6: Wire Stripping

Follow the directions to strip a wire.

❏ Check here when you've finished step 4.

7

Electrical Inventor's Journal

ideas for inventions • drawings • questions • experiments • problems • solutions • stories

Homemade Electric Devices

For each device, play the game, explore the device, then answer these questions:

1. What do you think you are supposed to do with it?

2. What are the parts you can see?

3. What parts do you think you **can't** see?

4. How do you think it works?

Station 7: Quiz Board

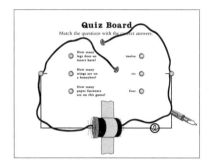 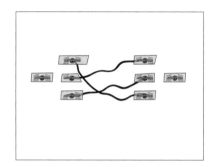

1. _____

2. _____

3. _____

4. _____

Station 8: Thread the Needle

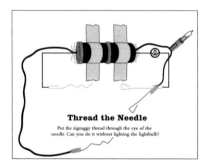

1. _____

2. _____

3. _____

4. _____

8

Electrical Inventor's Journal

ideas for inventions • drawings • questions • experiments • problems • solutions • stories

Answer the same four questions for whichever of the following stations are set up in your classroom.

Station 9: Ice Fishing

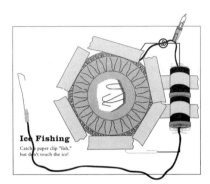

1. _____

2. _____

3. _____

4. _____

Station 10: Count Up

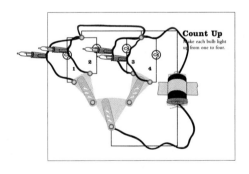 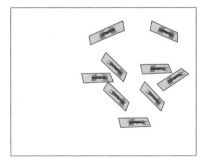

1. _____

2. _____

3. _____

4. _____

⑨

Electrical Inventor's Journal

ideas for inventions • drawings • questions • experiments • problems • solutions • stories

Station 11: What Matches?

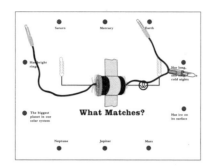

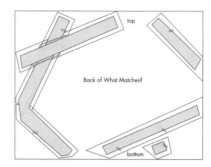

1. _____

2. _____

3. _____

4. _____

Station 12: Upstairs/Downstairs

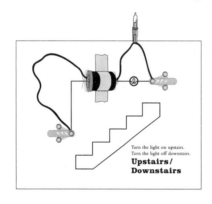

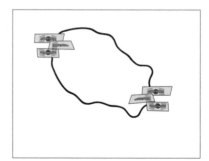

1. _____

2. _____

3. _____

4. _____

(10)

Electrical Inventor's Journal

ideas for inventions • drawings • questions • experiments • problems • solutions • stories

STUDENT HANDOUT
SESSION 2

Use this page for drawings or to write about any of these topics.

- **electric invention ideas**
- **stories of how we figured something out**
- **problems we had**
- **solutions we came up with**
- **experiments**
- **drawings**

11

Overview

After their explorations in the stations of the prior session, the students will be full of questions and ideas, and ready for some formal concept discussion. This session begins by introducing a variety of concepts that draw from the students' experiences at the stations. They then return to the same station setup as in Session 2, but each team goes to *stations it hasn't yet been to.* This time, as they visit each station, they have more time, a new perspective, and can apply the newly learned concepts to the stations.

First the class discusses the roles of the battery, conductors, and insulators in the electric devices. Then the entire class is challenged to make a holiday lightbulb light up—with every person involved, holding either a piece of wire, a lightbulb, or a battery to create a large class circuit. Once they're successful, they're introduced to the concepts of open and closed circuits. The students use a lightbulb as an indicator of whether a circuit is open or closed, and learn how to "follow the current" to find the path of a circuit.

Students then learn about how more voltage increases an electric current, and again discuss acceptable (and unacceptable) ways in which to experiment with electricity in school or at home. They're introduced to the symbols for battery, lightbulb, and wire or other conductor used in schematic diagrams. They'll use these symbols throughout the unit in diagramming a variety of circuits. Equipped with all these new concepts, the students set off once again to work with their partners at the learning stations.

Note: All students need to complete Stations 1 (Shocking Stories) and 2 (Build a Circuit) in order to proceed to Session 4, but not all students need to complete all the remaining stations.

"People Model of Electric Current," a recommended, optional model of how electric current flows at a subatomic level, appears as a Special Going Further at the end of Session 4. It includes all the concepts introduced to that point. You may want to do this with students after Session 4, or incorporate it into Sessions 3 and 4 as concepts are introduced. Read about the model on page 98 to help you decide if and how you'll present this session.

Key Concepts for Session 3

- Energy can be carried from one place to another by electric current.

- Electric circuits require a complete loop through which an electric current can pass.

Your students may have arrived at many of these concepts on their own during the previous session. You may choose to state or post one or more of them to emphasize.

- Electricity in circuits can produce light and heat.

- Energy can be stored in batteries.

- Electric current flows through some materials (conductors) and does not flow through other materials (insulators). Metals are electric conductors.

- When an electric circuit is open, the particles stop flowing and the current stops; when an electric circuit is closed, the current flows.

- More voltage makes more current flow through a circuit.

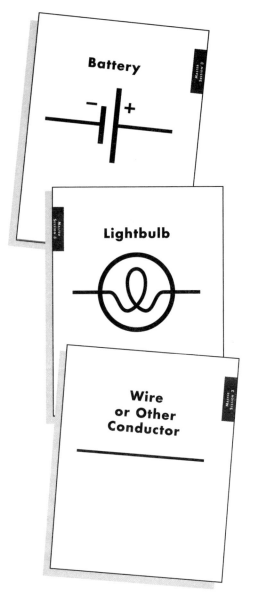

Battery

Lightbulb

Wire or Other Conductor

■ What You Need

For the class:
- ❑ the **No 120V** poster from previous sessions
- ❑ 1 copy each of the **Battery, Lightbulb,** and **Wire or Other Conductor** posters (pages 85–87)
- ❑ 1 holiday lightbulb with stripped leads
- ❑ 2 C-cell batteries
- ❑ the electric exploration stations, set up as in Session 2
- ❑ *(optional)* chart paper or sentence strips for writing out key concepts

For each student:
- ❑ the **Electrical Inventor's Journal** from the previous session
- ❑ 1 piece of 20-inch-long wire with stripped ends
- ❑ *(optional)* 1 different-colored pen for drawing circuits in the journal

■ Getting Ready

1. Make one copy each of the **Battery, Lightbulb,** and **Wire or Other Conductor** posters (pages 85–87). Have them ready to post on the wall.

2. Make sure all the materials and devices at the stations from Session 2 are in working order, and that the **No 120V** poster is still prominently displayed.

3. Either with the help of parent or student volunteers, or on your own, strip ¾ inch from both ends of the 20-inch-long wires. Each student will need one for the class circuit activity.

4. Set out the following materials:

- 1 piece of 20-inch-long wire with stripped ends for each student

- 1 holiday lightbulb with stripped leads

- 2 C-cell batteries

5. Set out the students' **Electrical Inventor's Journals** from the previous session. If you've chosen to have the students use a different-colored pen to draw the circuits in their journals, have them available.

6. If you've chosen to post one or more of the key concepts for this session (pages 73–74), write each one in large, bold letters on separate sheets of chart paper or sentence strips, and set aside to post later in the session.

■ How the Electric Devices Work

1. Have students think back to their travels through the stations in the previous session. Ask them what makes all the electric devices at the stations work. [The battery, which makes an *electric current* move through the device.]

2. Ask students to name objects that electric current flows through and objects they found it does *not* flow through, from the stations in the previous session. List these on the board. (Some students will have tested this directly at the Conductors and Insulators station, but even those who haven't been to that station yet should also be able to name some.)

The list will look something like this:

Things electric current flows through	**Things electric current does not flow through**
wire	cardboard
foil	tape
paper clip	paper
paper fastener	plastic film canister
aluminum or metal pie pan	
battery	
necklace	

3. If there's some disagreement over the list, borrow some material from the Conductors and Insulators station and ask one or two students to do a test in front of the class.

4. Ask the students what scientists call the materials the electric current flows through. [**Conductors.**] Write "Conductors" at the top of the first column on the board.

5. Ask if they can see anything the conductors have in common. [They're all (or mostly) made of some type of metal.]

6. Tell them a conductor is a material that allows electricity to flow through it. Some metals, the ground, and even water are good conductors. Add these to the list on the board.

7. Ask if they know what the materials that do not carry electric current are called. [**Insulators.**] Write "Insulators" at the top of the second column.

Conductors	**Insulators**
Things electric current flows through	**Things electric current does not flow through**
wire	cardboard
foil	tape
paper clip	paper
paper fastener	plastic film canister
aluminum or metal pie pan	
battery	
necklace	
the ground	
water	

8. Add "your body" to the list of conductors, because living beings can also conduct electricity. Say that if a strong electric current were to pass through a body, it could result in severe burns or even death. It's important for students to learn to protect themselves from coming into contact with electricity.

9. Remind them that the low voltage of the batteries they're using keep the activities safe.

10. Tell the students that at Station 3 they'll make their own list of conductors and insulators in their journals.

■ Creating a Class Circuit

1. Instruct the students to stand in one large circle and join with them if you'd like. Assign one student to hold a battery and a student on the opposite side of the circle to hold a holiday lightbulb.

2. Ask students what they would need to light the holiday lightbulb *without* bringing the lightbulb and battery together. [A long conductor.]

3. Pass out a 20-inch piece of wire with stripped ends to every student *except* the two students holding the lightbulb and the battery.

4. Ask students to connect the ends of their wires to the wires of their neighbors on either side, either by holding the wires together or by twisting them together. The students on either side of the student holding the battery should hold one end of their wires against one of the ends of the battery.

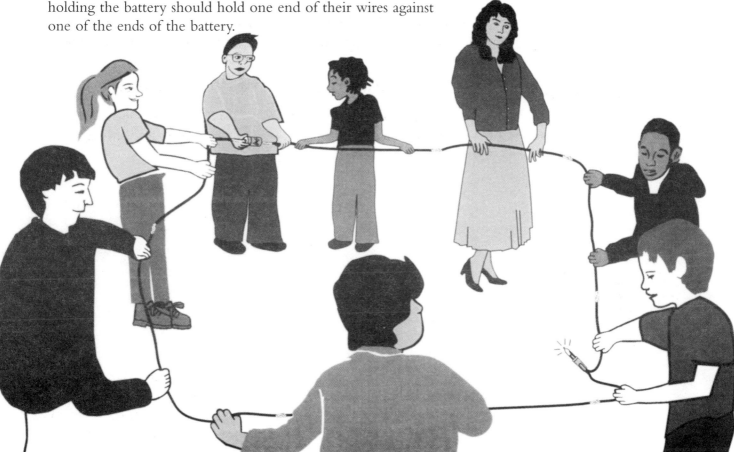

5. Emphasize that this activity requires teamwork. It may not be obvious to some students that it must be a metal-against-metal (conductor-against-conductor) connection between their wire and their neighbor's wire. Help them work together until the bulb finally lights up.

6. Point out that although the class has created a *large* circuit, from an electrical standpoint it's a "simple circuit." Say that they'll be learning more about this.

7. Keep the students in the circle for the following discussion.

Some teachers suggest using this activity at the beginning of the school year to teach cooperation and build teamwork. It can also be used as a "getting to know you" activity, in which each student gives her name and one fact about herself, then joins the circle. This continues until the circuit is complete and the lightbulb lights up.

■ Introducing Electric Current and Circuits

1. Explain to the class that the shining lightbulb shows there's an *electric current* flowing in the wire. Tell them they'll be doing more activities to learn how to control electric currents.

2. Tell them one way to understand what's happening in an electric device is to **follow the path the electric current takes.** "Following the current" will be an important strategy in this and other sessions in this unit.

3. With the students' help, follow the electric current. Starting from one end of the battery, follow the current through the set of wires in half of the circle, through the lightbulb (making it shine), through the remaining wires in the other half of the circle, then back to the battery.

4. Explain that the path of the current ends where it starts, somewhat like a circle (though it doesn't have to be circular). The name for the path of a current, a *circuit,* even sounds like the word "circle."

5. Tell students they can call it a circuit even when a connection is undone and the electric current doesn't flow. In that case, it's called an *open circuit.* When there is truly a complete circuit, with an unbroken path for the electric current to flow around, it's called a *closed circuit.*

6. Have the students open and close the circuit (which has probably been going on since they first succeeded in getting the lightbulb to shine). Help them identify whether the circuit is open or closed by whether or not the light is shining. You might note that any student has the power to open the circuit, but it takes all the students' cooperation to make a closed circuit.

7. Keep the students in the circle for more discussion.

■ More Voltage, More Current

1. Ask students how they could increase the current to make the light shine brighter. They'll probably suggest adding more batteries to the circuit.

2. Step into the circuit between two students with wires and insert another battery. Demonstrate that if the ends of the batteries with the bumps (the positive contacts) are facing the same direction around the circuit, the light *does* indeed shine brighter. The current *has* been increased. If, however, the batteries are pointing in opposite

directions in the circuit, they're working against each other, and no current flows.

3. Tell students that adding more batteries so they work together increases the power, or **voltage,** in the circuit.

4. Collect the equipment, thank the students for working well together, and have them return to their seats.

■ More Important Safety Messages

Voltage is an advanced concept, and teaching its true meaning is not within the scope of this guide. Voltage is introduced here to help identify electric-safety issues. It's enough that students know that the amount of voltage has something to do with the strength of the electricity.

1. Tell students that this activity helps illustrate an important message—circuits with **more than two batteries** (or with **different kinds of batteries**) should *never* have batteries installed pointing in opposite directions. Explain that two batteries could force the current to flow backward through a third battery that is backward in the circuit, causing an unwanted chemical reaction in the battery that could damage the battery or even cause it to explode.

2. Draw the class's attention to the **No 120V** poster, and review what it means:

   • Electricity from a wall socket is *never* going to be part of the electric circuits they build in this class.

   • Electricity from a wall socket must also *never* be used for experimenting with electricity at home.

3. Ask students who have visited the Shocking Stories station what they found out about what could happen to a person whose body acts as a conductor for 120 volts (or more). [The person could be injured or killed.]

4. Ask students who have been to the Shocking Stories station to share other safety tips they learned from reading the stories. If no one mentions the following, do so yourself:

   • Never touch power lines with anything, including kites and poles. Ask if they should try to get their kite or sneakers down if they're stuck in power lines. [NO. And they shouldn't fly kites anywhere near power lines or during lightening storms.]

   • Don't stick metal objects into electrical outlets.

   • Keep water and electricity apart.

   • Prevent short circuits (they'll learn about this concept later on).

- If you get a shock while using an electric appliance, unplug it immediately and report it to an adult, who should have the appliance repaired.

- If an electric appliance has loose wiring, don't use it.

- If an electric appliance is broken, don't repair it yourself—and never explore a device while it's still plugged in.

■ Schematic Diagrams

1. Tell students that when scientists need to describe their circuits for other scientists, they usually draw pictures. Many scientists have also been good artists (like Lewis Howard Latimer, who was a draftsman), but many are not, so scientists use a simple "picture language" to draw their circuits.

2. When scientists use this picture language, they make what's called a **schematic diagram.** This has simple pictures that don't look like the parts of the circuit, but do show how the parts of a circuit are connected.

3. Hold up a C-cell battery and draw on the board the schematic symbol for a battery.

4. Point out that the symbol doesn't look like a battery. Ask the students if there's anything about the picture that *is* like a battery. [It has two different ends for making connections, one negative, one positive. The "plus," or positive, end of the battery is the one with a bump on it.]

5. Above the symbol for a battery, draw the schematic symbol for a lightbulb. Again, ask students how the symbol is like a lightbulb.

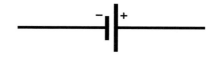

*What most people call a "battery" is, in fact, an electric **cell.** A true battery is more than one cell connected so that they work together. (See "Background for the Teacher.") Because the word battery is so commonly used, even by scientists, to describe cells, we've chosen to use "battery," but feel free to use the more accurate terms if you feel it's important.*

6. Call on a student to finish the diagram—connecting the symbols together to make a simple closed circuit. (Note that some students have trouble telling which part of the symbol for the battery the connections should be drawn to.)

7. Help the student interpret the symbols and accept advice from the other students until the schematic diagram for a simple closed circuit looks something like this:

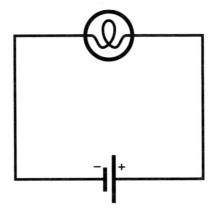

8. Point out to the students that the lines that connect the symbols in the schematic diagram represent any conductor. In the pictures, they're drawn thin and straight like wires—and they could be wires—but they could also be strips of foil, paper clips, pennies, or pie pans.

9. Post the **Battery, Lightbulb,** and **Wire or Other Conductor** posters. Let students know that for their reference these posters will remain up throughout the unit.

10. On the board, make a simple drawing of one of the homemade electric devices from Session 2.

11. Talk the students through the path the current takes through the circuit in the real device. For the Ice Fishing device, for example, say "the current goes from the non-bumpy end of one battery through both batteries and out the bumpy end, through the light, into the pie pan, then to the hook (when touched to the pie pan), and back to the non-bumpy end of the battery. This is the circuit."

■ The Circuit Hunt: More Electric Exploration Stations

1. Tell the students they'll circulate through the same stations as in Session 2, but now they'll be going to stations they haven't been to yet. (If you're adding more stations, briefly describe them.)

2. Let students know that another way this session is different from Session 2 is that they're going to be on a "circuit hunt." Let them know that as they go to each station (except Stations 1, 2, and 6) they need to be "hunting" for the circuit—by following the path of the current from the battery, through all the parts of the circuit, **and back to the battery.**

3. Tell students that in their journals they'll be using a different-colored pen (if you've provided this) to draw the circuits on top of the illustrations (those they've drawn and those already pictured). They'll do this both for stations they've already been to and for those they visit during this session.

You may choose to simply let your students freely explore the homemade electric devices without recording anything except their journal notes. You may also challenge them to record their answers to the questions at the stations (they may need to write on the back side or on extra paper) and try to draw the path of the circuit of each device (or just the easier ones) on each drawing.

4. Call students' attention to the simple drawing of the device on the board. With their help, use chalk (a different color, if possible) to trace the circuit of the device—starting at the non-bumpy end of the battery, going up through both batteries, through the light, into the pan, then on to the hook and back to the non-bumpy end of the lower battery—to make a closed loop.

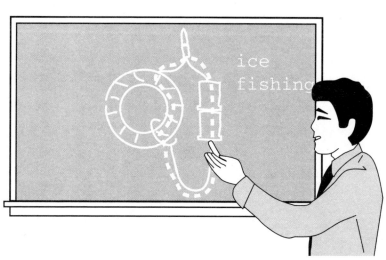

5. Now erase the chalk mark going through the battery and ask, "Is this still a drawing of a closed circuit?" [No.] Draw the line back in again and reiterate that the students will need to draw complete loops to represent how the current flows through each circuit.

6. Emphasize to students that if they haven't yet completed Stations 1 and 2 (Shocking Stories and Build a Circuit), they need to do those first, before doing any other stations.

7. Tell them that they may not get to all the stations, but that's OK; the electric devices will be available for them to explore throughout the unit.

8. Have students team up with the same partners they had during Session 2. Pass out the journals. Direct each team to a station it hasn't yet been to (or to Stations 1 and 2). Make sure the students are well distributed, then have them begin.

9. As teams work at the stations, circulate and check the following:

   • Are they all getting a chance to get their hands on the activities?

   • Have they completed Stations 1 and 2 before going on to others?

   • Are they thinking about the questions at the stations?

   • Are they looking at the homemade electric devices from the front and the back?

   • Are they using their **Electrical Inventor's Journals?**

   • Are they "following the current" and drawing it in their journals?

10. At the end of the class session as you're cleaning up the stations, leave available the homemade electric devices and the materials from Stations 2 and 5 for use in Session 4.

It's not necessary that every team complete every station. Although they should all have done Stations 1 (Shocking Stories) and 2 (Build a Circuit) by the end of this session, it's alright if they don't get to all the other stations—as long as they've had a chance to explore what does and doesn't make a circuit. If you feel it would be important for your students to have more time with the stations, you may choose to spend more time on them before the next session or at the beginning of the next session.

Energy can be carried from one place to another by electric current.

■ Optional: Key Concepts

1. If you've chosen to post some or all of the key concepts introduced in this session, hold them up one at a time, and have one or more students read them aloud. Briefly discuss how these statements review the important ideas from today's activities.

2. Post the concepts on the wall for students to refer to during the rest of the unit.

■ Going Further

Journal Writing

Give your students time to write and draw in their journals about what they've learned. Encourage them to include their discoveries, problems and solutions, ideas for future projects, and any questions they still have. Prompting students with specific examples can help make this more successful and rewarding. For example, "LaToya why don't you write about how you discovered your bracelet was a conductor?" Or "Jacob, you had some great questions about short circuits—why don't you write those down?"

Battery

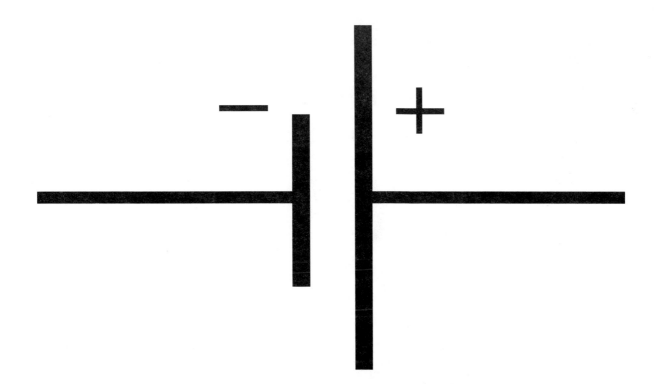

Lightbulb

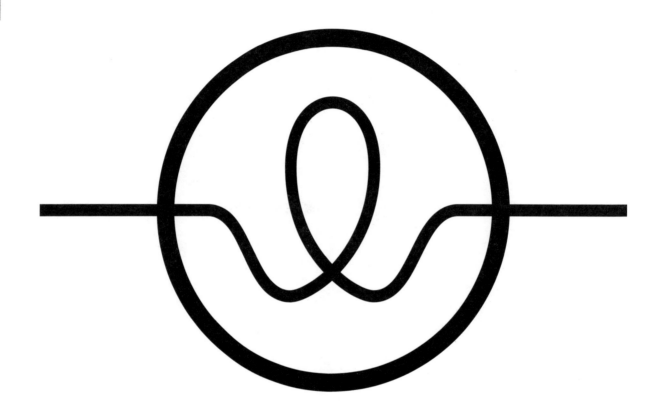

Wire
or Other
Conductor

Overview

This session begins with a debrief of the Build a Circuit station, focusing on the path of electric current through the circuit. The class then discusses the Light the Bulb station, providing another opportunity for students to grasp what does and doesn't make an electric circuit. If there's disagreement among the students as to which drawings worked (made the bulb light) at the station, they can be re-tested together as a class. Students go over each of the homemade electric devices, reinforcing their understanding of circuits. This also provides inspiration for future electric projects they may design, if you've chosen to do Sessions 10 and 11 with them.

Students are then introduced to the concept of a short circuit. They learn *not* to experiment with short circuits, and that if they make one by accident, they should immediately disconnect it.

In teams of four, students build each of the circuits depicted in schematic diagrams on the **Building Circuits from Diagrams** sheet. This activity helps them learn how circuits work and what will happen when things are added to or taken away from their circuits. It also gives practice using schematic symbols and diagrams, and will help them with their inventions later in the unit. In a teacher-led class debrief, these concepts are amplified by the students' experiences.

Key Concepts for Session 4

- In a short circuit, electric current flows very quickly, and may overheat the circuit while wearing down the battery quickly.

- A schematic diagram of a circuit can help in predicting how the circuit will work.

■ What You Need

For the class:
- [ ] the **No 120V, Battery, Lightbulb,** and **Wire or Other Conductor** posters from previous sessions
- [ ] 1 copy of the **No Short Circuits!** poster (page 105)
- [ ] 1 overhead transparency of the **Light the Bulb** station sign (page 55)

"People Model of Electric Current," the recommended optional model of how electric current flows at a subatomic level, appears as a Special Going Further at the end of this session (page 98). It includes all the concepts introduced to this point. You may want to do this with students at the end of the session, or you may have already chosen to incorporate it into Sessions 3 and 4 as concepts were introduced.

□ 1 overhead transparency of the **Building Circuits from Diagrams** sheets (pages 106–107)

□ a circuit device from the Build a Circuit station

□ a C-cell battery from the Build a Circuit station

□ a holiday lightbulb with stripped leads from the Light the Bulb station

□ the homemade electric devices used in Sessions 2 and 3

□ an overhead projector

□ *(optional)* chart paper or sentence strips for writing out key concepts

For each team of four students:

□ 1 strip from the **Experimenter's Kit List** (page 108)

□ 1 C-cell battery

□ 1 resealable plastic bag

□ 3 holiday lightbulbs with stripped leads

□ 4 pieces of 4-inch-long wire with stripped ends

□ 2 prepared film canisters and lids

□ 4 paper fasteners (brads)

For each student:

□ the **Electrical Inventor's Journal** from the previous session

□ 1 copy of the **Building Circuits from Diagrams** sheets (pages 106–107) to add to journal

■ Getting Ready

1. Make a copy of the **No Short Circuits!** poster (page 105). Re-post, if necessary, the **Battery, Lightbulb,** and **Wire or Other Conductor** posters from Session 3, and make sure the **No 120V** poster is still prominent.

2. Make one transparency each of the **Light the Bulb** station sign (page 55) and the **Building Circuits from Diagrams** sheets (pages 106–107).

3. Make two copies of the **Experimenter's Kit List** (page 108) so that when cut into strips you'll have one for each team of four students. (You may choose to laminate the strips for future use.) Cut them apart and have them ready to distribute to teams.

4. Set out the following materials for each team to put together into "Experimenter's Kits" in assembly-line fashion during class:

   • a resealable plastic bag

The "Experimenter's Kits" will also be used in Sessions 7, 8, and 9. To save time during class, you can assemble these kits ahead of time.

- 3 holiday lightbulbs with stripped leads (Not all three bulbs are needed for this activity, but will be when the kits are used in other sessions.)

- 4 pieces of 4-inch-long wire with stripped ends (Again, not all four wires are needed now, but will be in other sessions.)

- 2 prepared film canisters and lids

- 4 paper fasteners

5. Set out one C-cell battery for each team of four students.

6. From the materials used for the Build a Circuit and Light the Bulb stations in Sessions 2 and 3, gather a circuit device, a holiday lightbulb, and a battery. These materials will be used to debrief the stations. Later the circuit device and battery will be used to demonstrate a short circuit.

7. For their debrief, have available the homemade electric devices used in Sessions 2 and 3.

8. Make a copy of the **Building Circuits from Diagrams** sheets (pages 106–107) for each student. Have them ready to distribute along with the **Electrical Inventor's Journals** from the previous session.

9. If you've chosen to post the key concepts for this session (page 89), write them out in large, bold letters on separate sheets of chart paper or sentence strips, and set aside to post later in the session.

◼ Debriefing Stations 2 and 5: Build a Circuit and Light the Bulb

1. Debrief the Build a Circuit station:

   a. Hold up a circuit device made at the Build a Circuit station. (You needn't put a battery in it now.)

   b. Ask students to show where they think the electric current flows through the device, assisting if necessary. Remind them that it flows in a "circle" (loop).

2. Debrief the Light the Bulb station:

   a. Show the transparency of the **Light the Bulb** station sign. Ask students which drawing made the bulb light during their tests.

b. If they don't all agree that the drawing in the middle is the only one that works, use a holiday lightbulb and battery to have the whole class try out any others they think did light the bulb.

c. Ask volunteers to show where they think the electric current flows through the middle drawing. Again, remind them that it flows in a "circle" (loop). Ask students to look at the drawings in their journals. Tell them to trace the circuit there, if they haven't already.

■ Debriefing the Repair Shop

1. Ask students if they repaired something that wasn't working.

2. Ask if anyone has tips about repair, or any tricks for assembling things, that they'd like to pass on to their classmates.

■ Debriefing Stations 7–12: Homemade Electric Devices

1. As you debrief each of the stations as described below, hold up each homemade electric device and ask the students what they have to say about it.

2. During your debrief, take the opportunity to use terms introduced thus far.

Station 7: Quiz Board

Summary: On the back of the board, each question is connected by a wire conductor to its correct answer. When the question and answer are also connected in the front by the wires, a closed circuit is formed and the lightbulb lights up. When a question and a *wrong* answer are touched, it's not a closed circuit, because they're not connected on the back.

Station 8: Thread the Needle

Summary: When the two paper clips touch, a closed circuit is formed.

1. Ask the students to describe the path of the circuit.

2. Ask how this device is similar to Ice Fishing.

3. Ask what they could do to make the game harder. [Options include making the eye of the "needle" smaller or making the "thread" more "wiggly."]

Station 9: Ice Fishing

Summary: When the "hook" touches the "ice," a closed circuit is formed. The electric current can flow and the light shines.

Students may be reminded of the board game "Operation," in which players remove "bones" from a "patient."

Station 10: Count Up

Summary: The foil-covered paper clips are conductors, and can send the electric current to different lightbulbs.

Station 11: What Matches?

Summary: This one is a lot like the Quiz Board device, except it uses foil instead of wires as conductors to connect the correct matches on the back of the board. Also, there isn't a one-to-one set of correct matches.

Ask "Which hole is not connected to any other?" [Neptune.] "Which holes are connected to more than one other hole?" [Earth, Mars, and Has ice on its surface.] Students with some familiarity with electric circuits may realize that there must be an insulator between the strips of foil that cross but are not connected. This will be addressed in Session 7.

Station 12: Upstairs/Downstairs

Summary: Students can control the light from two places.

Ask if any of them have light switches at home that work that way. Are there any at the school?

■ Discussing Short Circuits

1. Draw the following schematic diagram on the board.

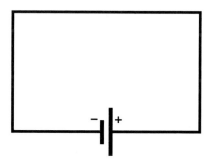

2. Ask students to follow the current with you as it leaves the battery, flows through the wire, and returns to the battery.

3. Ask students how this circuit is different from the other circuits they've used. [There's no lightbulb in the circuit.]

4. Tell students that when electricity can flow around a circuit *without* making a light shine—or a motor turn, or a toaster get hot; without making anything happen—it's called a **short circuit.**

5. Tell them the current in a short circuit can make the circuit get very hot. Assure them they can't get electrocuted from a C-cell battery, but the heat from a short circuit could cause discomfort or even a minor burn.

6. Ask students what kind of dangerous things could happen if there's a short circuit with a current that's 120 volts or stronger. They should all be able to share what they learned from the story about a short circuit at the Shocking Stories station.

7. If they don't mention the following, do so yourself: **short circuits can cause burns and fires, and can cause electric devices to stop working.**

8. Tell students that making short circuits is not only dangerous, but also wasteful. The strong current that flows in short circuits drains the battery quickly without accomplishing anything useful.

9. Tell them that in future activities they should try *not* to make short circuits, and that if they make them by accident they should disconnect them immediately.

10. Show students the **No Short Circuits!** poster then display it prominently near the other posters and refer to it regularly during the remainder of the unit.

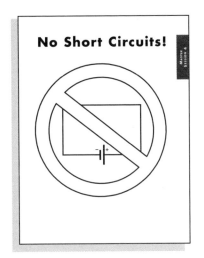

No Short Circuits!

■ Examples of Short Circuits

1. Recount these anecdotes about accidental short circuits:

   • A person puts a battery in a pants pocket that has keys and coins in it, and soon wonders why there's a hot spot on that side of her leg. (This happens most often with 9-volt batteries, which have the terminals close together. A 9-volt short circuit gets quite hot!)

   • When Thomas Edison was preparing to demonstrate his lightbulb to the press, someone placed a small piece of copper across the wires, trying to short circuit them so the demonstration wouldn't work. Luckily he was caught in the act, and the short circuit was prevented.

   • A person thinks the battery in his device will look better if it's wrapped in foil. Ask the class how that makes a short circuit.

2. Take one of the circuit devices students made at the Build a Circuit station and reinsert a C-cell battery into the film canister.

3. Without actually doing it, show that you could complete a circuit without touching the loose wires together:

   • You could stretch the wire with the lightbulb around to touch the paper fastener at the opposite end of the battery.

   • You could stretch the wire *without* the lightbulb around to touch the paper fastener at the opposite end of the battery.

4. Draw schematic diagrams of the two ways to complete the circuit, as shown below, and ask students which one shows a short circuit.

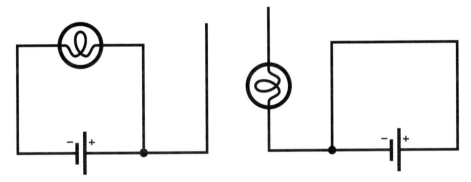

5. Point out how schematic diagrams help you discover problems with a circuit before you actually put it together. Schematic diagrams can also help show how a circuit works, or why it doesn't work after you've put it together.

■ Building Circuits from Diagrams

1. Show the **Building Circuits from Diagrams** transparencies and tell the students they'll have a chance to build each of the different kinds of circuits shown in the diagrams (except circuit C).

2. Explain that this will help them learn how circuits work, and will give them practice using the symbols and diagrams. It will also help them with their inventions later on.

3. Go through each of the steps on the transparencies, making sure your students understand them.

   • Look at the circuit diagram.

   • Write what you predict will happen when you build your circuit (predict *before* you build it, and don't change your prediction once it's built!).

You may want to tell your students to write their predictions in pen, to discourage them from changing them after the fact.

- Build the circuit.

- Write what happened.

- Try to explain why it happened.

4. Emphasize that in the second and fifth steps (predicting and explaining), a "right" answer is not the point. Let students know the point is to help them understand circuits better.

5. Be sure to point out the following:

- In circuit C, they will *not* build the circuit, but will still write a prediction and an explanation of why they think their prediction would be correct. [The circuit in C is a short circuit, which is why they won't be building it. Don't tell them this now—you'll want to know who does and doesn't figure it out unaided.]

- In circuit G, students will record how many bulbs and batteries they plan to use, write their prediction, **build a circuit of their own design,** draw it using schematic diagram symbols, then write what happened and why.

6. Divide students into teams of four. Give each team four copies of the **Building Circuits from Diagrams** sheets (one per student), one strip from the **Experimenter's Kit List,** and one C-cell battery. (If their plan for circuit G involves two batteries, you'll need to provide an extra one to the team.) Pass out the **Electrical Inventor's Journals** from the previous session, and ask students to add the **Building Circuits from Diagrams** pages to it.

7. Make sure students understand that although they are working as a team to build the circuits, they will individually fill out the **Building Circuits from Diagrams** sheets in their journals.

8. Tell the students to assemble their "Experimenter's Kits" by gathering the listed items from the designated area and putting them in the resealable plastic bag. Allow a few minutes for this. (Check that students don't take more than their share of materials.)

9. Let students know that before they build their circuits, they must first make a battery holder. Point out the illustration at the top of the **Building Circuits from Diagrams** transparency and have students find it on their journal page. Then give them the following instructions:

a. Insert a paper fastener into the slit in the bottom of the film canister. The "head" of the fastener should remain on the bottom of the canister.

Some students' first attempts at making diagrams will probably combine the schematic symbols with a more "true-to-life" attempt at illustration. During the next few sessions, watch the development of their ability to make simple and accurate diagrams.

b. Reach into the canister, separate the legs of the paper fastener, push them apart, and press them firmly against the inside of the bottom of the canister.

c. Insert another paper fastener into the lid of the film canister from the inside to the outside. Separate the legs of the fastener and push them firmly against the top of the lid.

d. Insert the C-cell battery into the film canister with the bump-side (positive end) down and the flat side (negative end) up.

e. Fasten the lid on the canister. You will have to push hard around the edge of the lid so that it stays on firmly.

10. When teams have assembled their kits and made a battery holder, they can begin working. Check to see that every student is engaged in the circuit building and that all are recording information in their journals.

11. Ten or 15 minutes before the end of class, have students remove the batteries from their circuits and reassemble their "Experimenter's Kits" by placing the listed items and the **Experimenter's Kit List** strip in the plastic bag. Collect the batteries and kits. Have students keep their journals to refer to during the class discussion.

■ Debriefing the Circuit-Building Activity

1. Tell your students one of the purposes of the activity was to help prepare them for making circuits later in the unit. The activity helps give ideas of what will happen when different things are added to or taken away from their circuits. Ask the following questions:

   • What did you need to make a circuit that made the bulb light? [Battery, lightbulb, conductor, and closed circuit.]

   • What happened when you had a lightbulb in the circuit, but no battery (circuit B)? [No light.]

   • What did you predict would have happened with a battery in the circuit, but no lightbulb (circuit C)? [Eek! Short circuit!]

   • What happens when there's a short circuit in real life? [The battery wears down quickly, gets hot, and—if there's enough voltage—can start a fire.]

The response to the question about circuit E mentions "particles" (in reference to electrons). Depending on your students' developmental level, you may choose to refrain from introducing the concept of particles, and instead simply use the word "current."

- What happened to a circuit when a second lightbulb was added (circuit E)? [The bulbs were dimmer. The particles/current slowed down.]

- What did you notice about the other circuits?

2. Ask students to comment about how circuit connections were made, how many bulbs each circuit had, and how brightly they shined.

3. If you're going to do Sessions 10 and 11, remind students that they'll eventually be designing their own devices and should record any ideas they came up with during this session.

4. Let them know that in the next session, they'll be inventing switches.

5. If you've chosen to post the key concepts introduced in this session, hold them up and have a student read them aloud. Briefly discuss how each reviews the important ideas from today's activities. Post the concept on the wall for students to revisit during the rest of the unit.

■ Going Further

1. Journal Writing
Give your students time to write and draw in their journals about what they've learned, their discoveries, problems and solutions, ideas for future projects, and unanswered questions.

2. Creative Circuits
If you've chosen to teach only Part I of this unit, you may want to allow your students some time to design and assemble their own creative circuits at this point. These can be as simple as adding another lightbulb, paper clip, or piece of foil to the simple circuit, and can provide a fun and simple challenge. They can then use these circuits in their circuit design in the following session.

■ Special Going Further
People Model of Electric Current

Simple Circuits and Short Circuits

This model of how electric current flows at a subatomic level includes the following topics related to Sessions 1 through 4:

Please be sure to read the What Is Electricity? and Misconceptions about Electricity sections of the "Background for the Teacher" section (page 176) before you present this Special Going Further activity.

- simple circuits

- no battery

- short circuit

- no bulb

- two bulbs

- their own inventions

What You Need

- ☐ 1 **clear** incandescent (not fluorescent) lightbulb (filament visible)
- ☐ a C-cell battery
- ☐ 8–10 pieces of 8 ½ inch x 11 inch paper for signs
- ☐ 1 or 2 sponges
- ☐ a paper cutter
- ☐ 2 small boxes
- ☐ 1 tube approximately 5 inches in diameter and 8 inches long
- ☐ masking tape

Getting Ready

1. Label the two small boxes "battery" to represent the battery in the model. Any box will do, but the closer to a battery shape it is, the better. Label one end of each box with a plus symbol and the other ends with a minus symbol to represent the positive and negative terminals of a battery.

2. Using the paper cutter, cut up the sponge(s) into approximately 40 pieces to represent electrons. Put them in the "battery" boxes.

3. In large, bold letters, write the words "Off," "Dim," "Bright," and "Very Bright" on four separate pieces of paper. At different times during the activity, you'll hold these signs up near the "lightbulb" student to show how brightly, or not, the light is shining.

4. Make the following signs on strips of paper. These will be taped to students to identify their role in the model.

   - 1 that reads "lightbulb"

   - 2 that read "battery"

   - 3–5 that read either "wire" or "atoms in the wire"
     (See the Electrons section on the next page to help you decide which term to use.)

5. The large tube will be used by the lightbulb student to pass or roll sponge pieces through during the activity. It can be anything big

Depending on your students' levels of concept understanding, you may choose either to advise them on how to depict each aspect of the model, or challenge them to come up with the components themselves.

The shape of the lids of large "red vines" candy containers (often available in warehouse grocery stores) is like the bumpy-end of a battery. These containers can be used as the battery model.

If you choose to do the second optional people model activity, in Session 9, you'll need one extra battery box and two extra tubes. You'll also need more small signs to label students— two more "lightbulb" signs, one more "battery" sign, and a new "switch" sign. You may wish to make the extra materials now and set them aside.

enough for the student to reach through, including a soda bottle or milk carton with the top and bottom cut off, or a rolled-up piece of cardboard. Prepare the tube, label it as the filament, then set it out along with the clear lightbulb, the battery, and the rest of the material you've just prepared.

1. Model of a Simple Circuit

Electrons

1. Tell the class you're going to ask for several volunteers to act out a model of what's going on inside circuits, to demonstrate the nature of electric current. Tell them **electric current is made up of flowing electrons.**

2. First, explain that everything—all matter—is made up of *atoms.* Atoms are so tiny they can't be seen with ordinary microscopes.

3. Now say that *electrons* are even smaller than atoms. Some electrons are stuck to atoms. Other electrons move freely between and around the atoms in a conductor. Let students know these "freely moving" electrons are what they will demonstrate in the model.

4. Select six to ten students to use as volunteers for the model, and ask them to stand in a row at the front of the room. Using the signs you made earlier, label one of the students "battery" and another "light-bulb." Label the remaining students as "wire" or "atoms in the wire," whichever you prefer. (Alternatively, you can have *all* the students in your class participate as wires or atoms.)

5. Hold up a few of the sponge pieces. Tell students that since real electrons are far too small to see, in this model you'll pretend these sponge pieces are electrons. Give each "wire" student one "electron."

6. Ask your class how they think the electrons will move in the model when the current is started. Listen to and validate all their ideas, then tell them that when the current begins to flow they'll be passing electrons to each other in a particular direction. Remind them that these are the electrons that move freely among the atoms in the wire.

Battery

1. Hold up one of the boxes with sponge pieces, and tell the students it represents a battery.

A deep understanding of atoms is not necessary for this model, though if you think your students are ready for it, you may choose to expand in that direction. If not, you can still refer to the particles as electrons, and focus on the flow of current without your students knowing electrons can be components of atoms.

*If you choose to make the model more accurate, give each "wire" student a few sponge "electrons" to keep in her pocket during the activity. This represents the fact that the atoms do hold on to their own electrons. The flowing electrons that make up the electric current are **additional** electrons, moving between and around the atoms.*

2. Hold up the real C-cell battery and say that a real battery pulls electrons in from one end and pushes electrons out at the other end. Remind students that inside this *model* battery (the box), the sponge pieces represent electrons. Ask what the voltage would be in the model if it represents a regular C-cell battery. [Between one and two volts.]

3. Tell students that even though the battery is always pushing the electrons, they don't actually go anywhere until there's a complete closed circuit.

4. Give the box of sponge pieces to the battery student. Explain that during the activity, a wire student will be putting electrons into the box, and the battery student will be taking them out.

Lightbulb

1. Hold up the clear lightbulb and point out the filament inside. Tell students the very thin filament wire is harder for the electrons to get through than the other wires in the circuit. Remind the class that when developing the lightbulb Edison's team tested thousands of filament materials to find one that would both glow brightly and last a long time.

2. Hand the lightbulb student the tube and tell the class it represents the filament. Whenever current is flowing, the lightbulb student needs to pass the electrons to the next person by reaching through the tube.

3. Show the four pieces of paper with "Off," "Dim," "Bright," and "Very Bright" written on them to the class. Tell the lightbulb student that part of her job will be to decide which one of these signs should be displayed, depending on the electron flow. Also tell her that during the activity she should call out the "brightness," and you will display the appropriate sign. Say that, for instance, at normal flow, she should call out "bright," and you will display that sign, leaving her hands free to pass electrons through the filament.

Flowing Current

1. Ask all the volunteers what they think they should do when the current is not flowing. [No electrons should be passed, and the lightbulb student should call out for you to display the "Off" sign.]

2. Hold up the real C-cell battery again, and point out that batteries have two ends, or **terminals.** Show the students that the battery box is marked with a positive and negative end.

3. Ask the class how the students acting as wire, lightbulb, and battery would need to be connected to make an electric circuit.

4. Have the volunteers arrange themselves. Verify and explain that they're in a circle so they represent wire going from one terminal of the battery to the lightbulb, then through more wire back to the battery's other terminal.

5. Give an electron (sponge piece) to the lightbulb student. Make sure the wire students understand that their sponge pieces represent electrons already inside the metal of the wire they represent. The electrons are **not** from the battery.

6. Tell them that now that they're "connected," electrons will be taken from the battery at the negative (non-bumpy) end and passed into the wire there. The electrons will be passed through that wire, through the lightbulb, through more wire, and back to the positive (bumpy) end of the battery.

7. Explain that, furthermore, the wire student next to the positive end of the battery will drop electrons into the battery box, and the battery student will pick up electrons from the box to pass them on to the wire student at the negative end.

8. Tell the volunteers they will pass the electrons on as they receive them, but to **always make sure they have one in their hand.** All the electrons in the students' hands should move together. Students should give electrons away with one hand while receiving them with the other. This takes teamwork and practice. Congratulate them when they finally succeed.

9. To start the demonstration, ask the battery student to begin handing electrons, one at a time, to the wire student at the negative end of the battery. Once the "current" is moving well ask the lightbulb student which sign you should show. Hold up the Bright sign above her so everyone can see it.

10. After the students have got the hang of it, and as the model continues flowing, ask the class to focus on one electron you select, and together follow it around the circuit, describing each step. (You might quickly spray or apply a splash of color on that piece, to make it easy to follow.) One way to make this step fun is to use a horse race announcer's voice, "…she's comin' down the wire, slipping between those atoms…."

Open and Closed Circuits

1. Now announce that you are an item made of rubber (an insulator), and step between two wire students to break the circuit. All the students should immediately stop passing electrons.

2. Ask your students what the model now represents. [An open circuit.] Step away and put the wire students back together. The students should immediately start passing electrons again. Ask what the model represents. [A closed circuit.]

3. Use this same method of making an open circuit (and halting the activity) every time you wish to make a point when doing the remainder of the activity.

Adding More Batteries

1. Now ask the class how you could increase the electric current—the number of electrons that move through a wire in a given amount of time. [Add another battery.]

2. Hold up the other battery box. Ask if it matters where in the circuit it is placed. [No.] Ask if it matters what direction the terminals are pointed. [Yes; the second battery's terminals must be pointed in the same direction as the first battery's.]

3. Ask another student to volunteer to be a second battery, label him, and give him the second battery box. Add the new battery student to the circuit. Make sure the volunteers understand that **all the electrons in the circuit must now move more quickly.**

4. Ask the class what other effect having two batteries may have on the circuit or ask the lightbulb student what she should call out. Hold up the Very Bright sign.

5. Ask what's making the bulb brighter. [After adding the second battery, there is more voltage.] Ask what the voltage in the circuit is now. [About three volts.]

6. Ask the volunteer wires: "Did you feel yourselves getting warmer as the current was increased?" [When the current increases, a real wire will heat up too. That's the way electric stoves, toasters, and hair dryers work.]

2. Model of a Short Circuit

1. Open the circuit and remove the lightbulb student from the model. Ask the class what they think will happen when the circuit is closed again.

2. Close the circuit and ask students what they notice. [The electrons should be passing around the circuit more quickly, because there's less **resistance.** They'll learn more about this in Session 7.]

3. Ask if they think this could cause any problems. [Yes, it could run down the battery more quickly and overheat the wire.]

4. Ask what this kind of circuit is called. [Short circuit.] Remind students **never** to deliberately set up short circuits, and that if they do by accident, to disconnect them immediately.

5. Thank the volunteers and collect the materials.

No Short Circuits!

ideas for inventions • drawings • questions • experiments • problems • solutions • stories

Building Circuits from Diagrams

The battery holder looks like this:

You put it together like this:

In the schematic diagram you draw it like this:

A

Look at the circuit diagram.

Write down what you predict the circuit will do.

Build the circuit.

Write down what the circuit did.

Try to explain why the circuit did that.

B

Look at the circuit diagram.

Write down what you predict the circuit will do.

Build the circuit.

Write down what the circuit did.

Try to explain why the circuit did that.

C

Look at the circuit diagram.

Write down what you predict the circuit would do.

Do not build this circuit.

Why should you **not** build this circuit?

Explain why you think your prediction is correct.

⑫

Electrical Inventor's Journal

ideas for inventions • drawings • questions • experiments • problems • solutions • stories

D

Look at the circuit diagram.

Write down what you predict the circuit will do.

Build the circuit.

Write down what the circuit did.

Try to explain why the circuit did that.

E

Look at the circuit diagram.

Write down what you predict the circuit will do.

Build the circuit.

Write down what the circuit did.

Try to explain why the circuit did that.

F

Look at the circuit diagram.

Write down what you predict the circuit will do.

Build the circuit.

Write down what the circuit did.

Try to explain why the circuit did that.

G

Draw your own circuit diagram, showing the bulbs and batteries you plan to use.

Write down what you predict the circuit will do.

Build the circuit.

Write down what the circuit did.

Try to explain why the circuit did that.

13

Experimenter's Kit List

Put these in the bag:

❏ 3 holiday lightbulbs

❏ 4 pieces of wire

❏ 2 film canisters and lids

❏ 4 paper fasteners

Keep this out of the bag:

❏ 1 C-cell battery

Experimenter's Kit List

Put these in the bag:

❏ 3 holiday lightbulbs

❏ 4 pieces of wire

❏ 2 film canisters and lids

❏ 4 paper fasteners

Keep this out of the bag:

❏ 1 C-cell battery

Experimenter's Kit List

Put these in the bag:

❏ 3 holiday lightbulbs

❏ 4 pieces of wire

❏ 2 film canisters and lids

❏ 4 paper fasteners

Keep this out of the bag:

❏ 1 C-cell battery

Experimenter's Kit List

Put these in the bag:

❏ 3 holiday lightbulbs

❏ 4 pieces of wire

❏ 2 film canisters and lids

❏ 4 paper fasteners

Keep this out of the bag:

❏ 1 C-cell battery

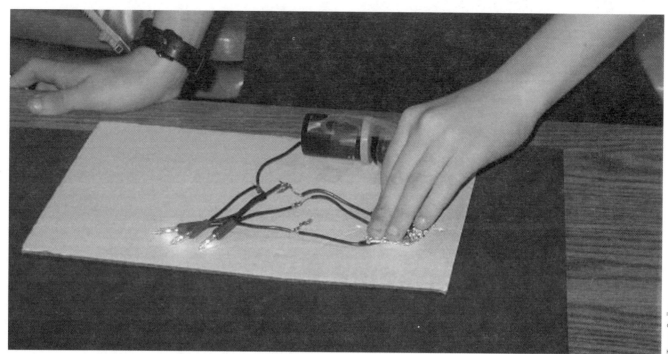

The three things that are essential to achievement are hard work, stick-to-it-iv-ness, and common sense. —Thomas Edison

Overview

At this point in the unit some of your students may be "champing at the bit" to begin their own electric inventions. In this session they get their first taste, by inventing something relatively simple and specific— a switch. Though all will benefit from the experience of inventing, some of your students will, of course, have an easier time than others in coming up with an idea, having patience when things don't work out quite as they envisioned, and with general *"stick-to-it-iv-ness."*

The session begins using a simple circuit to introduce the idea of a switch and how students can use a switch to make a circuit open or closed. In keeping with previous sessions, they're introduced to schematic diagram symbols for open and closed switches.

Next, the class brainstorms devices they know about that use switches, and note the wide variety of switches. They're challenged in teams of two to invent their own switch. You may be amazed at how many different kinds of creative switches your students come up with and how engrossing the activity can be for them. They write up their switch inventions on **Switch Invention Spec Sheets.** They'll share their inventions with the rest of the class in the next session, as well as get a chance to try out their classmates' inventions.

■ What You Need

For the class:

- ❑ the **No 120V, Battery, Lightbulb, Wire or Other Conductor,** and **No Short Circuits!** posters from previous sessions
- ❑ 1 copy of the **Switch** poster (page 119)
- ❑ 1 roll of aluminum foil
- ❑ an assortment of metal paper clips, large and small
- ❑ several paper fasteners (brads)
- ❑ several pieces of wire with stripped ends
- ❑ extra wire to cut and strip as needed
- ❑ several pairs of wire cutters and strippers
- ❑ several scraps of cardboard, of various thicknesses

"My students were making switches unknowingly during the previous stations by just playing around with the materials they had. The connection was much easier to grasp."
—Cheryl Lemke, trial-test teacher

The more materials students have to work with, the more creative they can be. Consider this a basic list, and feel free to include other materials not listed here.

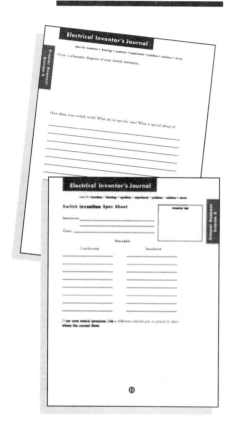

For each pair of students:
- ❏ 1 C-cell battery
- ❏ a circuit device from the Build a Circuit station
- ❏ masking tape or stickers

For each student:
- ❏ the **Electrical Inventor's Journal** from the previous session
- ❏ 1 copy of the **Switch Invention Spec Sheet** (pages 117–118) to add to journal

■ Getting Ready

1. Make a copy of the **Switch** poster (page 119). Have it ready to display with the posters from the previous sessions.

2. Gather the circuit devices and batteries from the Build a Circuit station. You should have one of each for each pair of students.

3. Also gather tape or stickers. Each pair of students will use this to label their switches and spare parts with their names. Gather the pieces of wire students stripped at Station 6 in Sessions 2 and 3.

4. Select a part of the room to be the supply area. Stock it with cardboard pieces, extra paper clips and fasteners, foil, wire pieces, etc.

5. Make one copy of the **Switch Invention Spec Sheet** (pages 117–118) for each student.

6. Set out students' **Electrical Inventor's Journals** from previous sessions.

7. Plan to allow a few minutes of time at the end of the session for students to discuss their switches.

■ Introducing Switches

1. Hold up one of the circuit devices students made at the Build a Circuit station. Insert a battery and touch the wires together to close the circuit and light the bulb.

2. On the board, draw a schematic diagram for a simple closed circuit. Review each part of the diagram (battery, lightbulb, and wire) with students.

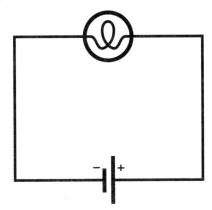

3. Point out that most of the circuits people use in their everyday lives have something that makes them closed circuits when the flow of electric current is wanted, and makes them open circuits to stop the flow of current. In other words, most circuits have **switches** to turn them on and off.

4. Put the **Switch** poster on the wall with the others. Add a switch symbol to the schematic diagram on the board.

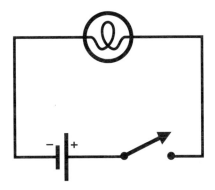

5. Open and close the switch in the drawing on the board, pointing out that when it's "on," the circuit is closed and the current can flow. When the switch is "off," the circuit is open and no current flows.

6. Hold up the actual circuit device again, pointing out and demonstrating that when the light is on, the circuit is closed, and when the light is off, the circuit is open.

7. Use one or more electric switches in your classroom (such as a light switch, an overhead projector, or a flashlight) to illustrate the same concept; on means a closed circuit, off means an open circuit.

8. Have your students list devices that have switches. Discuss the variety of ways in which switches work:

- there are switches you flip

- switches you slide

- switches with a knob to twist

- switches with a button to push

- switches with a chain to pull

- switches that stay "on" when you turn them on

- switches that go "on" only when you press them and "off" when you stop pressing—a doorbell, for instance

- switches that go "off" when pressed and "on" when they're released—refrigerator lights are an example

- switches that turn other switches "on" and "off"; computers have millions of tiny switches that do this

■ Inventing Switches

1. Look again at the actual circuit device and point out that the loose wires make a switch, but it's crude and hard-to-use. Tell the class they'll work in pairs to invent their own kind of switch.

2. Let the class know that designing and building the switch is just the first part of the job. After they've done that they must each make a "spec" sheet. Explain that *spec* is short for *specifications,* and a specifications sheet is a paper that has important information about an invention. In some ways it's like an advertisement—most inventors want people to buy and use their invention.

3. Add that if the students were building a device that needed a switch, they'd look at the spec sheets of several different switches from several companies to help decide which switch would work best for their device. The spec sheet alone would tell them what they needed to know about the switch.

4. Show students the **Switch Invention Spec Sheet** journal page and say they'll each fill one out. Tell them their spec sheets must include the following:

- the names of the inventors

- the date

- a list of conducting materials in the circuit

- a list of other materials used (insulators)

- a drawing of the switch, including the path where current flows

- a schematic diagram of the switch

- a description of how the switch works and any specific uses for the switch

Let students know their spec sheets could also include:

- a company logo

- a description of why the switch is of especially high quality

- other illustrations

5. Tell them their spec sheets should be clear enough that someone could use it as a "recipe" to build the switch without ever seeing the actual switch.

6. Show students the supply area and point out the materials they have to work with. Tell them that after they've planned their switch they can go there to get the material they need.

7. Distribute a **Switch Invention Spec Sheet** journal page to each student.

8. Give the student pairs a few minutes to plan their switch before you distribute the batteries and circuit devices from the Build a Circuit station. Give each pair the tape or stickers with which to label their switches and any spare parts with their names, then allow them to begin construction.

9. As students work, monitor their projects and the amount of material they use. Remind pairs, if necessary, that building a switch doesn't require inventing a new kind of circuit. The simple circuit must still be at the heart of each invention.

10. When most students seem to have finished, call on a few to share some of the problems they encountered and what they did to try to solve them.

■ Cleaning Up

1. Fifteen minutes before the end of class, warn students that they'll be stopping soon, but that they'll have a few minutes in the next session to wrap up their inventions and spec sheets if they haven't finished. Ten minutes before the end, have them stop working and

tell them to label their switches and any spare parts. Ask students to remove the batteries from their switch inventions. Collect the batteries.

2. Ask the students to turn in parts they don't plan to use, and throw out or recycle any scraps of unusable materials.

3. Have them add the **Switch Invention Spec Sheet** to their **Electrical Inventor's Journals.**

■ Going Further

1. Journal Writing

Give your students time to write and draw in their journals about what they've learned, their discoveries, problems and solutions, ideas for future projects, and unanswered questions they still have.

2. Advertisement Posters

Have students design posters advertising their switches. This can be a great opportunity to practice persuasive writing skills and demonstrate their knowledge of their devices.

Electrical Inventor's Journal

ideas for inventions • drawings • questions • experiments • problems • solutions • stories

Switch Invention Spec Sheet

Inventors:_____

_____

Date:_____

Materials

Conductors

_____

_____

_____

_____

_____

_____

_____

Insulators

_____

_____

_____

_____

_____

_____

_____

Draw your switch invention. Use a different-colored pen or pencil to show where the current flows.

STUDENT HANDOUT SESSION 5

© 2004 The Regents of the University of California. May be duplicated for classroom or workshop use.

LHS GEMS • Electric Circuits

Session 5 **117**

Electrical Inventor's Journal

ideas for inventions • drawings • questions • experiments • problems • solutions • stories

STUDENT HANDOUT
SESSION 5

Draw a schematic diagram of your switch invention.

How does your switch work? What are its specific uses? What is special about it?

15

Switch

SESSION 6: ELECTRIC SWITCH TRADE SHOW

The ability to foresee, visualize, and create your idea in your mind is the most powerful resource you have. —George Washington Carver

Overview

This session provides all the student pairs the chance to share their switch inventions with fellow team inventors. The session begins with an opportunity for pairs to put "finishing touches" on their inventions and spec sheets to make sure they're ready to share with their classmates. Each pair is then divided into the roles of "Rover" and "Explainer." The Rover's job is to find out about all the switches his classmates have made, and how they work. The Explainer's job is to tell other students about her team's switch and answer any questions they may have. The Explainer is also in charge of repairs the team's switch may need during the session. After approximately 10 minutes, the two partners switch roles.

■ What You Need

For the class:
- ❏ the **No 120V, Battery, Lightbulb, Wire or Other Conductor, No Short Circuits!,** and **Switch** posters from previous sessions
- ❏ the extra material (aluminum foil, metal paper clips, paper fasteners, extra wires, etc.) from Session 5

For each pair of students:
- ❏ 1 C-cell battery
- ❏ the circuit device with switch they invented in Session 5
- ❏ *(optional)* 1 piece of paper labeled as "Rover's Recording Sheet"

For each student:
- ❏ the **Electrical Inventor's Journal** from the previous session

■ Getting Ready

1. Set out the batteries, switch inventions, and the extra materials from Session 5 so students can finish their work.

2. Set out the **Electrical Inventor's Journals** from the previous sessions.

3. If you'd like students to use it, write "Rover's Recording Sheet" at the top of a piece of paper and copy enough for each pair of students.

GO ■ **Finishing Touches**

1. Explain to your class that "trade shows" are big professional gatherings where companies go to show off their products—to each other, and to other companies that might want to buy them. Let students know that in this session, teams will have an Electric Switch Trade Show of their own, to share their switch inventions with their classmates.

2. Say that before the trade show starts, each pair will need a finished switch and their individual spec sheets for the switch.

3. Give pairs about 15 minutes to finish their switches and spec sheets, and to clean up.

■ The Trade Show

1. Explain that during the trade show, everyone should leave their spec sheets with their switch inventions. Divide each pair into a Rover and an Explainer. Tell the students they'll switch roles halfway through the activity.

2. Tell the Rovers they'll be visiting other inventors in the trade-show exhibit hall (the classroom). Their job is to find out as much as they can about the different switches on display. If you made "Rover's Recording Sheets," distribute one to each pair of students.

3. Say that the Explainer will stay by the team's switch to answer questions posed by visiting Rovers and demonstrate how the switch works. The Explainers also get to make on-the-spot repairs.

4. Point out that it will be the Explainer's choice whether to let visiting Rovers operate the switch themselves. If a switch is likely to come apart, the Explainer may choose to simply demonstrate how it works.

5. If you plan to do Sessions 10 and 11 (Electric Inventions and Electric Inventions Festival), tell students that as they look at the different switches designed by their classmates they can get ideas for electric inventions they will be making later in the unit. Let students know they should take their journals along and write down these ideas as they travel through the trade show.

6. Let Rovers rove for about 10 minutes, then have the partners change roles and go about 10 minutes more.

To one teacher it seemed that the students hardly understood the project at the end of the first day. When they picked up on the next day, however, there was a big change. They seemed to focus in on exactly what was needed to finish the projects.

■ Going Further

Journal Writing

Give your students time to write and draw in their journals about
what they've learned, their discoveries, problems and solutions, ideas
for future projects, and unanswered questions they still have.

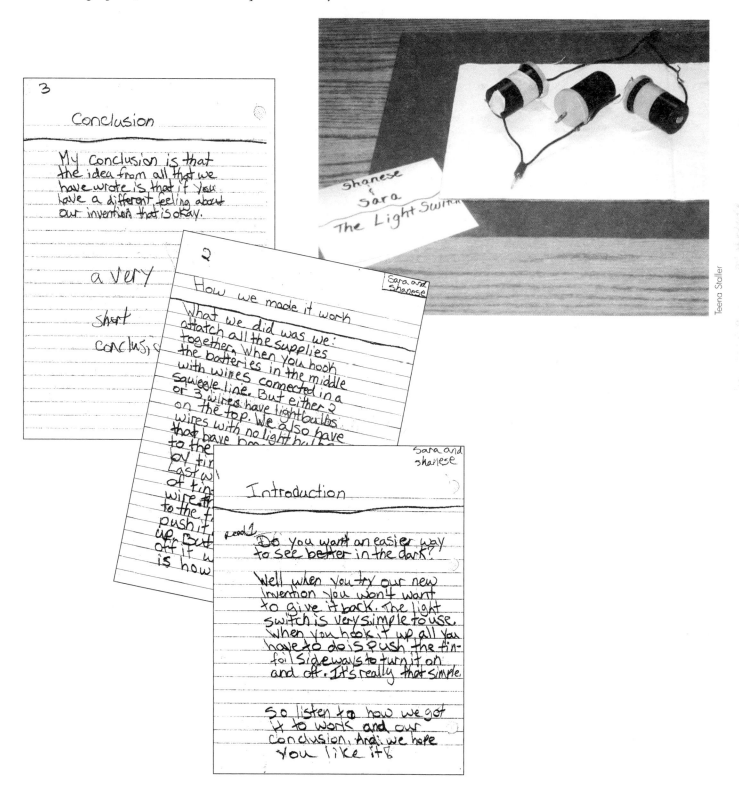

3

Conclusion

My conclusion is that
the idea from all that we
have wrote is that if you
have a different feeling about
our invention that is okay.

a very

short

conclusi

2

How we made it work

What we did was we:
attatch all the supplies
together. When you hook
the batteries in the middle
with wires connected in a
squiggle line. But either 2
or 3 wires have lightbulbs
on the top. We also have
wires with no lightbulbs
that have bee...
to the
o' tir
Last w
of tim
wire
to the
push it
up. But
off if u
is how

Sara and
Shanese

Introduction

Do you want an easier way
to see better in the dark?

Well when you try our new
invention you won't want
to give it back. The light
switch is very simple to use.
When you hook it up all you
have to do is push the tin-
foil sideways to turn it on
and off. It's really that simple.

So listen to how we got
it to work and our
conclusion. And we hope
You like it

Shanese
&
Sara
The Light Switch

Teena Staller

Part II

More Circuits
and More About
Circuits

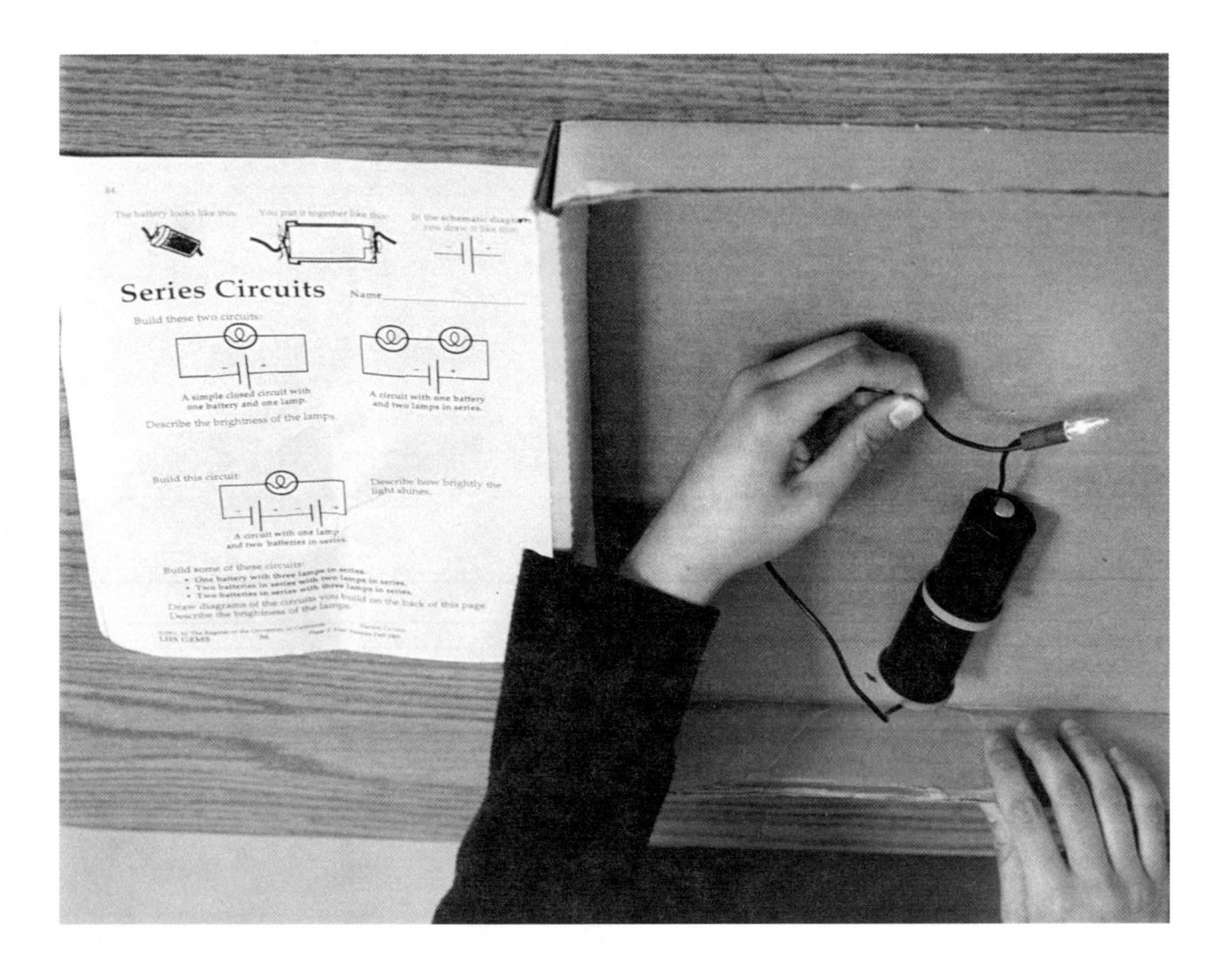

Overview

This session introduces students to series circuits and the concept of electric resistance. They begin in pairs, looking at two schematic diagrams of circuits and predicting which will have the brightest lights. In teams of four, students then build and test several circuits. They discover that when they add more lightbulbs in series, the current decreases and the lightbulbs glow more dimly. From this they also determine that the brighter the light, the more current is flowing; the dimmer the light, the less current. If your students did the People Model of Electric Current, the Special Going Further at the end of Session 4, they'll remember this phenomenon. Students see that putting batteries together to increase the voltage is another example of a connection in series.

The concept of resistance is explained, and examples are given of a circuit with very little resistance (a short circuit) and too much resistance (an open circuit). In addition to the concepts of resistance and series circuits, through the activity students also get more practice in building circuits that match simple schematic diagrams. They extend this experience by devising circuits with more components in series, and creating the schematic diagrams themselves.

The People Model "Special Going Further" activity at the end of Session 9, in the same way as the model at the end of Session 4, can be broken up into shorter vignettes to incorporate into Sessions 7, 8, and 9 as each concept is introduced. See page 149 to decide if and how you'll present this optional activity.

Key Concepts for Session 7

- A *series circuit* is a circuit in which the current flows through each component (such as a lightbulb) one after the other.

- When electric current flows through something like a lightbulb filament, there's more resistance. The current is slowed, and energy (such as light and heat) is given off.

- In a circuit, the path with the least resistance will carry the most electric current.

- A short circuit is a circuit with very little resistance.

- When current must travel through more-resistant materials (such as more than one lightbulb in a series circuit), less current flows, and the battery runs down more slowly.

■ What You Need

For the class:

❑ the **No 120V, Battery, Lightbulb, Wire or Other Conductor, No Short Circuits!,** and **Switch** posters from previous sessions

❑ extra and replacement parts
___ wire with stripped ends
___ wire strippers
___ holiday lightbulbs with stripped leads
___ paper fasteners
___ prepared film canisters and lids

❑ *(optional)* chart paper or sentence strips for writing out key concepts

For each team of four students:

❑ 2 C-cell batteries
❑ 1 "Experimenter's Kit" from Session 4

For each student:

❑ the **Electrical Inventor's Journal** from the previous session
❑ 1 copy of the **Series Circuits** sheet (page 133) to add to journal

■ Getting Ready

1. Make one copy of the **Series Circuits** sheet (page 133) for each student.

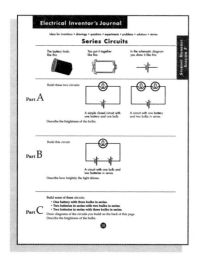

2. Set out the **Electrical Inventor's Journals** from previous sessions.

3. Set out one "Experimenter's Kit" and two C-cell batteries for each team of four students. Note that the "Experimenter's Kit" normally contains just one battery. In this session, however, students will need an additional battery.

4. Have available extra and replacement stripped wires and lightbulbs, as well as paper fasteners and prepared film canisters. Also have available one or more wire strippers.

5. Be sure all the posters put up in previous sessions are still prominently displayed.

6. If you've chosen to post one or more of the key concepts for this session (page 127), write each one in large, bold letters on separate sheets of chart paper or sentence strips, and set aside to post later in the session.

▶ ■ Building Series Circuits

1. Tell students that in today's activity they're going to start putting together circuits with more than one lightbulb.

2. Pass out the **Electrical Inventor's Journals** from the previous sessions. Also pass out a copy of the **Series Circuits** sheet to each student and ask them to add it to their journals. Organize the students into pairs and have them look at the first two circuits on the page. Ask them to try to predict which one will have brighter lights.

3. After a minute or so, tell students that next they'll be building the **first two circuits** in teams of four. Also, let them know that although they'll work as a team, each student will write observations on her own journal page. Ask students to refer to pages 12 and 13 of their journals to remind them of the circuits they built from diagrams in an earlier session.

4. Point out the reminder at the top of the **Series Circuits** sheet about how the battery holder goes together, but say that the students may use any method that works to make connections to the battery.

5. Tell students that **after they've made the first two circuits,** they may try the others shown and listed on the sheet. Point out that they'll need to take the first two circuits apart and combine the parts in order to have enough materials to build the other circuits.

6. Emphasize that if a team builds circuits that don't already have diagrams, the students should **draw their own diagrams** on the back of the sheet, and write a description of how brightly the lights shine.

7. Remind them to **avoid making short circuits** and point to the **No Short Circuits!** poster.

8. Organize students into teams of four and pass out one "Experimenter's Kit" and two C-cell batteries to each team. Let students know that you have extra material available for them should they need it. Tell them to begin.

9. After about 20 minutes, when each team has built four or five circuits, have students remove the batteries from their circuits and hand them in.

10. Allow the teams to keep their circuits intact (but without batteries) for demonstration purposes during the discussion—but either tell them to set their circuits aside or conduct the discussion in another location, so students won't be tempted to fidget with them.

■ Analyzing Series Circuits

1. Draw the circuit with one battery and two lightbulbs in series on the board.

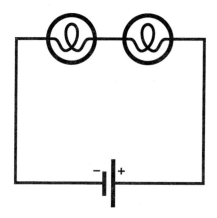

2. With the class, follow the current in the circuit and comment that the electric current flows through "one bulb after another." Explain that *series* is a word that means "one after another." Tell students that "television series" or "baseball series" are some ways people use the word. Let students know the circuits they built in this session are called ***series circuits.***

3. Have volunteers describe the difference in how lightbulbs shine in a circuit with one bulb and a circuit with two bulbs in series—the first two circuits on the journal page. [Two bulbs in a series circuit glow much more dimly.]

4. Draw a schematic diagram of the circuit with one lightbulb next to the circuit with two lightbulbs.

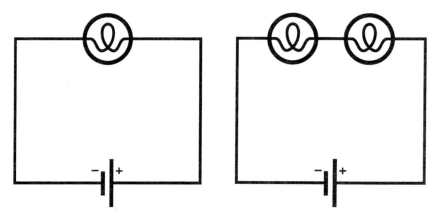

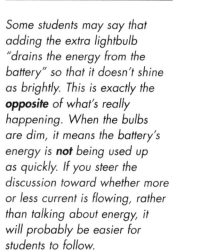

Some students may say that adding the extra lightbulb "drains the energy from the battery" so that it doesn't shine as brightly. This is exactly the **opposite** *of what's really happening. When the bulbs are dim, it means the battery's energy is* **not** *being used up as quickly. If you steer the discussion toward whether more or less current is flowing, rather than talking about energy, it will probably be easier for students to follow.*

5. Ask students how they might guess which circuit has more electric current flowing in it. Confirm that the brighter the light, the more current. A dimmer light means there's less current.

6. Tell them the conductor in a lightbulb is a long, thin wire called the ***filament.*** The filament is tightly coiled—it's longer than it looks! Let students know that it is harder for current to flow through such a long, thin wire than to flow through shorter, thicker wire. Remind the students that when developing the lightbulb, Edison's team tested thousands of filament materials to find one that would both glow brightly and last a long time.

7. With the class watching, again follow the current around each of the circuits drawn on the board, but this time, whenever you get to a lightbulb, act as if it takes real effort for the electric current to push through. (Gritting your teeth, scrunching up your face, and grunting usually help get the point across.)

8. Tell students that anything that reduces the flow of current has **resistance.** Each lightbulb adds more resistance to the circuit. If the circuits had buzzers or motors or heaters in them, these would also add resistance.

9. Say that, while the resistance of the filament in the bulb reduces the flow of current, it's also what makes the lightbulb shine. The current flowing through the filament heats the wire up so much that it glows brightly. The more current, the more brightly the filament glows.

10. Find out whether any teams built a circuit with one battery and **three lightbulbs in series.** Ask what they noticed. Confirm that the lights were very dim, or didn't shine at all. If a team has such a circuit at their desks or tables, you can hand them a battery and have them demonstrate.

11. Ask students to recall what an open circuit is. Tell them that an **open circuit** has so much resistance that no current can flow. Ask them what kind of circuit has almost no resistance. [A short circuit!]

12. Say that in a short circuit there's little resistance to slow the current, and the battery will be drained quickly and heat will be generated.

13. Ask why the lights in a short circuit don't shine extra bright. Confirm that they don't because the current doesn't flow through lightbulbs in a short circuit!

14. Tell students that with one battery, more resistance results in less current. Ask them if they think there's a way to increase the current, even with more resistance. [Add more batteries.] Ask which teams made series circuits with two batteries. Give them two batteries and have them demonstrate their circuit for the class.

To help explain the concept of resistance, have the class imagine that an electric current is like a line of students walking around the perimeter of the classroom. Then have them imagine that something happens to make the path very narrow— desks too close to the wall, for example. What happens to the current? [Since the line of students (the "current") has to pass through a narrow path, fewer students can pass through. The narrow opening creates resistance to their movement.]

Most students benefit from frequent repetition and reinforcement of the concept of resistance. The more you can incorporate it into your discussions and descriptions, the better.

15. Ask students to disassemble their circuits and put the materials and the **Experimenter's Kit List** in their "Experimenter's Kit" bags. Collect these and the journals, and tell students that during the next session, they'll put circuits with more than one bulb together in a different way.

■ Optional: Key Concepts

1. If you've chosen to post some or all of the key concepts introduced in this session, hold them up one at a time, and have one or more students read them aloud. Briefly discuss how these statements review the important ideas from today's activities.

2. Post the concepts on the wall for students to refer to during the rest of the unit.

■ Going Further

Journal Writing

Give your students time to write and draw in their journals about what they've learned, their discoveries, problems and solutions, ideas for future projects, and unanswered questions they still have.

Electrical Inventor's Journal

ideas for inventions • drawings • questions • experiments • problems • solutions • stories

Series Circuits

The battery looks like this:

You put it together like this:

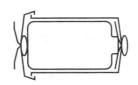

In the schematic diagram you draw it like this:

Part A

Build these two circuits:

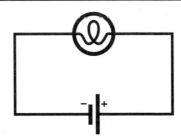

A simple closed circuit with one battery and one bulb.

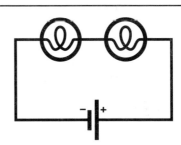

A circuit with one battery and two bulbs in series.

Describe the brightness of the bulbs.

Part B

Build this circuit:

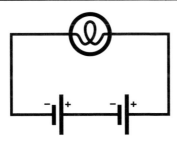

A circuit with one bulb and two batteries in series.

Describe how brightly the light shines.

Part C

Build some of these circuits:
- **One battery with three bulbs in series.**
- **Two batteries in series with two bulbs in series.**
- **Two batteries in series with three bulbs in series.**

Draw diagrams of the circuits you build on the back of this page.
Describe the brightness of the bulbs.

Overview

In this session, students compare two types of circuits: circuits with lightbulbs in series, with which they're familiar, and circuits with lightbulbs in parallel. They begin by following the current in schematic diagrams of a series circuit and a parallel circuit, noticing the "branch" in the parallel circuit. They learn the schematic diagram symbols for "connection" and "no connection"—symbols that indicate whether or not crossing conductors are connected.

In teams of four, students build the two circuits for themselves, and see that the lightbulbs in the parallel circuit shine more brightly than those in the series circuit. Students also notice that when they remove a lightbulb from the series circuit, the other lightbulb goes out, but when they remove one from the parallel circuit, the other bulb continues to shine.

The class is then introduced to the concept of *parallel circuits,* and learn that the reason the lightbulb shines more brightly in the parallel circuit is that the current splits up, or *branches*—and on each branch, encounters the resistance of only one bulb. In contrast, the current flowing in the series circuit goes through (and is slowed down by) the resistance of every bulb. This is a somewhat tricky concept. The more straightforward lesson from this session is that removing a bulb from a series circuit breaks the whole circuit, while removing a bulb in parallel doesn't affect the current in the other parts of the circuit.

After building and discussing simple series and parallel circuits, students have the chance to design more complex circuits and practice making schematic diagrams.

Key Concepts for Session 8

- A *parallel circuit* is a circuit in which the current splits and runs through separate components.

- In a parallel circuit, electric current divides into different pathways. The combined pathway has less resistance, allowing more current to flow and wearing the battery down more quickly.

■ What You Need

For the class:

❑ the **No 120V, Battery, Lightbulb, Wire or Other Conductor, No Short Circuits!,** and **Switch** posters from previous sessions

❑ 1 copy of the **Connection/No Connection** poster (page 143)

❑ the Quiz Board and What Matches? devices from Sessions 2, 3, and 4

❑ extra and replacement parts
   __ wire with stripped ends
   __ wire strippers
   __ holiday lightbulbs with stripped leads
   __ paper fasteners
   __ prepared film canisters and lids

❑ *(optional)* chart paper or sentence strips for writing out key concepts

For each team of four students:

❑ 1 C-cell battery

❑ 1 "Experimenter's Kit" from Session 4

For each student:

❑ the **Electrical Inventor's Journal** from the previous session

❑ 1 copy of the **Parallel Circuits** sheet (page 142) to add to journal

■ Getting Ready

1. Make one copy of the **Connection/No Connection** poster (page 143). Have it ready to display with the posters from the previous sessions.

2. Make one copy of the **Parallel Circuits** sheet (page 142) for each student.

3. Set out the **Electrical Inventor's Journals** from previous sessions.

4. Set out one "Experimenter's Kit" and one C-cell battery for each team of four students.

5. Have available extra and replacement stripped wires and lightbulbs, as well as paper fasteners and prepared film canisters. Also have available one or more wire strippers.

6. Gather the Quiz Board and What Matches? (if you made it) devices from Sessions 2, 3, and 4.

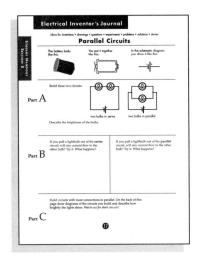

7. Be sure all the posters put up in previous sessions are still prominently displayed.

8. If you've chosen to post one or more of the key concepts for this session (page 135), write each one in large, bold letters on separate sheets of chart paper or sentence strips, and set aside to post later in the session.

■ Series Circuit and Parallel Circuit: Comparing Schematic Diagrams

1. Pass out the **Electrical Inventor's Journals** from the previous sessions. Also pass out a copy of the **Parallel Circuits** sheet to each student and ask them to add it to their journals. Point out the reminder about the battery holder at the top of the page, then ask the students to follow the current, starting from the battery and ending at the battery, in each of the schematic diagrams.

2. Draw the students' attention to the schematic diagram of the circuit labeled "two bulbs in parallel," and ask them to tell you what it has that the other circuits they've built did not have. [There's a branch in the circuit where the current can go in two different directions.]

3. Point out that where two or more conductors are connected, the schematic diagram shows a large spot. Tell students the spot is the symbol used to show that there's a connection that current can flow across.

4. Hold up the Quiz Board and What Matches? (if you made it) devices and say they're examples of circuits that have **conductors that cross but do not connect.** Show students the back of the devices, where the conductors cross. For the What Matches? device you may have to point out that there's tape between the strips of foil that cross.

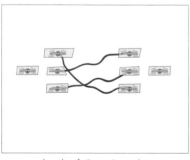

back of Quiz Board

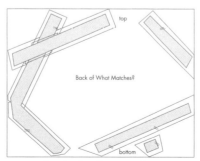

back of What Matches?

5. Put up the **Connection/No Connection** poster. Point out the symbol at the top and show students the large spot indicating that the crossing wires are connected. Also show students that wires that cross or come together but are **not** connected are represented by the symbol that looks like an arch. Point out the symbol on the bottom of the poster.

6. Tell students that whenever they draw a schematic diagram showing wires (or other conductors) that come together, they should use one of these two symbols (the spot or the arch) to make clear whether there's an electric connection or not. Let students know the poster will be up to remind them of the symbols.

■ Building the Series and Parallel Circuits

1. Either have students work in the same teams of four as in the previous session, or organize them into new teams. Tell the teams they'll build each of the circuits shown in part A of the **Parallel Circuits** sheet. Remind students that each of them will record observations on his own journal page. Point out that they'll need to take apart the first circuit and combine parts in order to have enough material for the second circuit, so they'll need to be sure to describe the brightness right after they build the first circuit.

2. Be sure students notice that after they build the circuits in part A, the directions in part B instruct them to **remove a lightbulb from each circuit** to see what happens.

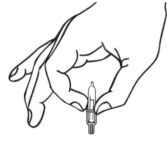

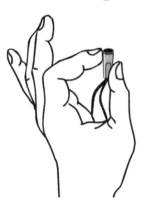

The concept of parallel circuits is a difficult one for most students to grasp. This activity has been carefully scaffolded to help them attain it. After completing parts A and B, students stop to participate in a class discussion. After the discussion, they are more prepared to go on to the final part of the student sheet. Some students may "catch on" to parallel circuits right away. They can move on to part C when they're ready— even if it's before the class discussion.

3. Demonstrate how to remove a bulb by prying with your fingernails or pulling the small rim of plastic at the base of the bulb with one hand, while holding the socket with the other. **Don't pull on the wires or the glass bulb itself.** (*Note:* It's easier to get the bulb back in after the demonstration if you don't pull it all the way out. If you do this, explain to students that they WILL pull the bulb all the way out.)

4. Tell the teams that if they finish parts A and B before the rest of the class, they may work ahead to do part C. Assure them that if they don't get to it now, they will be given time after the class discussion.

5. Remind students to **avoid making short circuits** and point to the **No Short Circuits!** poster.

6. Pass out a C-cell battery and an "Experimenter's Kit" to each team of four students, and let them work on their circuits. Let them know that as in the previous session, you have extra material available.

7. Circulate and help teams follow the diagrams accurately. Often, students will make series circuits while attempting to make parallel circuits. If they complete a series circuit while trying to make a parallel circuit, have them compare the finished circuit to each diagram to see which one it matches.

8. When all teams have completed parts A and B, collect the batteries and prepare to start a class discussion.

9. As in the previous session, allow the students to keep their circuits intact, without batteries, for demonstration purposes during the class discussion, but continue in such a way that students won't be tempted to fidget with them.

■ Comparing Parallel and Series Circuits

1. On the board, draw the schematic diagrams for the two circuits students built.

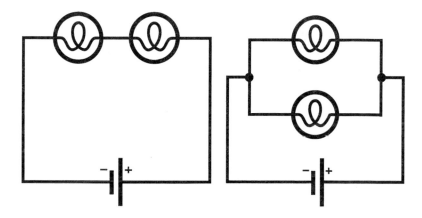

2. Ask students to describe the brightness of the lightbulbs in each circuit. [They were very dim in the series circuit but brighter in the other circuit.]

3. Ask students to recall what they learned from their experiences with series circuits to try to explain why the bulbs in series were dim. [The resistance of the filament in the two lightbulbs allows only a little electric current to flow in a circuit with one battery.]

4. Point out that in a series circuit there's a single path for current to flow in.

5. Ask students what ideas they have about why the lightbulbs in the other circuit were brighter. Guide them in their explanations, if necessary, by having them follow the current around the circuit. Help them see that in the second circuit, the current goes two ways—and either way the current goes, it flows through only *one* lightbulb.

6. You and the students may have been using the word "parallel" already to refer to the second circuit. Now is the time to formally introduce them to the concept. Tell students that a circuit that has a place where the current splits, or branches, is called a ***parallel circuit.*** Point out that in a parallel circuit there is more than one path for current to flow in.

7. Ask students about the word "parallel" and where they've heard it used before. It's likely that some of them will know—and share with you—its geometrical meaning, referring to lines that are side-by-side and never meet.

8. Tell students that a parallel *circuit* is different from parallel *lines,* because although there are parts in a parallel circuit where electric current flows separately through different conductors, eventually the "side-by-side" conductors *always* meet. In fact, they meet in more than one place. Point this out on the schematic diagram on the board.

9. To review part B of the **Parallel Circuits** sheet, erase a bulb in each schematic diagram on the board to simulate the removal of a bulb from each circuit. Call on students to explain their thoughts about why the remaining bulb in the series circuit goes out while the remaining bulb in the parallel circuit stays on. If necessary, have students "follow the current" around each circuit to help in their explanation.

10. Ask your students if they think the different lightbulbs in your classroom (or their homes or any other building) are in a series circuit or a parallel circuit. Ask them if they've ever noticed a burned-out lightbulb. Do the other lights still work? [Yes. Most circuits in buildings are parallel.]

■ More Explorations

1. Have a team (or two) that has worked ahead and built a more complicated circuit reinsert the C-cell battery and demonstrate the circuit for the class. Ask them to draw a schematic diagram on the board and tell the class how they think their circuit works.

2. Tell the students there are many other circuits they can build with the materials in their "Experimenter's Kits."

   • They can put three lightbulbs in parallel.

   • They can put one bulb in series with two other bulbs that are in parallel with each other, or two bulbs that are in series in parallel with a third lightbulb.

3. Tell them that just saying all that is probably harder than building the circuits themselves. Show them the diagrams *if you wish,* or leave them the challenge of drawing them in their journals.

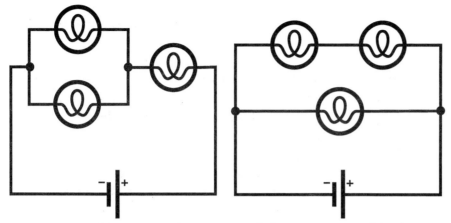

Two bulbs in parallel, in series with a third bulb.

Two bulbs in series, in parallel with a third bulb.

4. Ask students what kind of circuit they must not make. [Short circuits!] Tell them that as circuits get more complex, it's easier to make a short circuit by accident. Remind them to be alert and to disconnect short circuits immediately.

5. Redistribute the C-cell batteries to all teams. Refer students to the instructions on the bottom of their journal page, and let them work on part C for the remainder of the session.

6. Just before the end of the session you may wish to let some teams show their parallel circuits and draw the schematic diagrams on the board.

7. At the end of the session, have teams disassemble their circuits and put the materials back into their "Experimenter's Kit" bags. Collect these, the batteries, and the journals.

8. If you've chosen to post the key concepts introduced in this session, hold them up and have a student read them aloud. Briefly discuss how each reviews the important ideas from today's activities. Post the concept on the wall for students to revisit during the rest of the unit.

"I spent another session and required them to make and draw at least one circuit using three bulbs that had series and parallel on the same circuit. This solidified the difference between the two."
—Teena Staller, trial-test teacher

■ Going Further

Journal Writing
Give your students time to write and draw in their journals about what they've learned, their discoveries, problems and solutions, ideas for future projects, and unanswered questions they still have.

Electrical Inventor's Journal

ideas for inventions • drawings • questions • experiments • problems • solutions • stories

Parallel Circuits

The battery looks like this:	You put it together like this:	In the schematic diagram you draw it like this:
	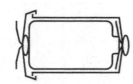	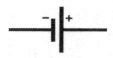

Part A

Build these two circuits:

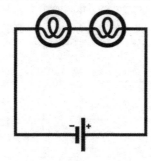 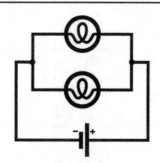

Describe the brightness of the bulbs in the two circuits.

Part B

If you pull a lightbulb out of the **series** circuit, will any current flow to the other bulb? Try it. What happens?	If you pull a lightbulb out of the **parallel** circuit, will any current flow to the other bulb? Try it. What happens?

Part C

Build circuits with more connections in parallel. On the back of this page draw diagrams of the circuits you build and describe how brightly the lights shine. *Watch out for short circuits!*

Connection

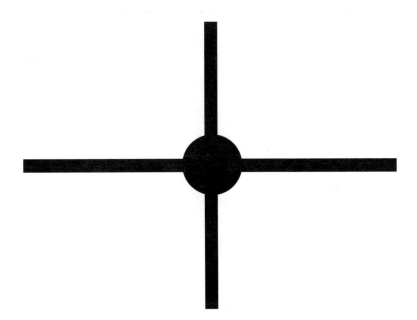

No Connection

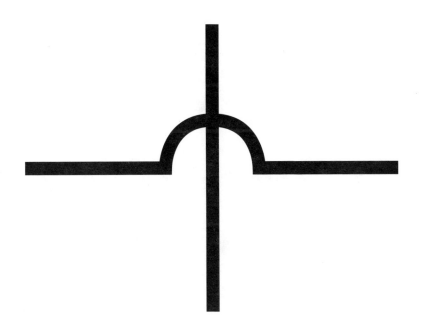

Overview

In this session, each team of four students is given a sheet with several challenges to redraw schematic diagrams and build new circuits. The teacher assigns each team to one challenge, and teams that are successful with that one may then take on another. Although at this point they're all veteran switch inventors, for this exercise students are told to simply use wires that they connect and disconnect to act as switches.

The challenges are examples of series and parallel circuits, some of which involve switches that turn one light off and others that turn two off. Students' attempts to make all the lightbulbs, or just certain light-bulbs, go on and off give them practice in predicting what a circuit will do, and reinforce the concepts of series and parallel circuits.

The second "People Model of Electric Current" (much like the one at the end of Session 4) is a Special Going Further. This time the model includes the topics related to Sessions 7 through 9.

■ What You Need

For the class:
- [ ] the **No 120V, Battery, Lightbulb, Wire or Other Conductor, No Short Circuits!, Switch,** and **Connection/No Connection** posters from previous sessions
- [ ] extra and replacement parts
  - __ wire with stripped ends
  - __ wire strippers
  - __ holiday lightbulbs with stripped leads
  - __ paper fasteners
  - __ prepared film canisters and lids

For each team of four students:
- [ ] 1 C-cell battery
- [ ] 1 "Experimenter's Kit" from Session 4

For each student:
- [ ] the **Electrical Inventor's Journal** from the previous session
- [ ] 1 copy of the **Switches in Series and Parallel Circuits** sheet (page 158) to add to journal

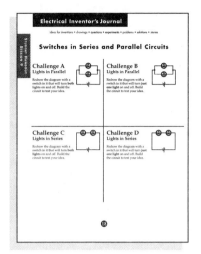

■ Getting Ready

1. Make one copy of the **Switches in Series and Parallel Circuits** sheet (page 158) for each student.

2. Set out the **Electrical Inventor's Journals** from previous sessions.

3. Set out one "Experimenter's Kit" and one C-cell battery for each team of four students.

4. Have available extra and replacement stripped wires and lightbulbs, as well as paper fasteners and prepared film canisters. Also have available one or more wire strippers.

5. Be sure all the posters put up in previous sessions are still prominently displayed.

■ Exploring Switches

1. Pass out the **Electrical Inventor's Journals** from the previous sessions. Also pass out a copy of the **Switches in Series and Parallel Circuits** sheet to each student and ask them to add it to their journals.

2. Point to the **Switch** poster to remind students what the symbol looks like. Let them know they'll be using this symbol in the schematic diagrams they'll be drawing in their journals.

3. Either have students work in the same teams as in previous sessions, or organize them into new teams. Assign a challenge from the sheet to each team. (More than one team may be assigned to the same challenge.) Remind them that although they'll work on the challenge as a team, each student will redraw the schematic diagram in her own journal.

4. Tell the teams that if they're successful at their assigned challenge, they may choose a new challenge from the sheet.

5. Tell the students they don't have to make fancy switches. They should simply use wires they connect and disconnect to act as switches.

6. Let students know that as in previous sessions, you have extra material available.

7. Remind students to **avoid making short circuits** and to immediately disconnect them if they make one accidentally. Point out the **No Short Circuits!** poster.

8. Pass out a C-cell battery and an "Experimenter's Kit" to each team of four students, and let them work on their challenge(s) for about 10 minutes.

■ Discussing Switches

1. Call on teams to describe and demonstrate how they tackled the challenges. Help them draw their schematic diagrams on the board and have them describe what happened with their circuits. Remind them to "follow the current," starting from the battery, to help describe what the circuit does. Have students demonstrate how the light(s) go on and off using the circuits they built. (It's not necessary that the class come up with solutions to all the challenges. The goal is to get all the students participating in a discussion of how electric current flows in a circuit.)

2. Here are some descriptions of how the challenges can be solved. In most cases, there are several places to put a switch that will do the job.

Challenge A: Turning both lights on/off in a parallel circuit

A switch in series with the pair of lights that are parallel with each other will turn them both off.

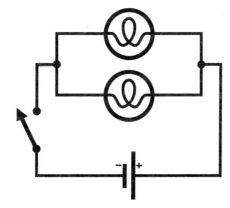

Challenge B: Turning one light on/off in a parallel circuit

A switch in series with one light, which together are parallel with another light, will turn off just one.

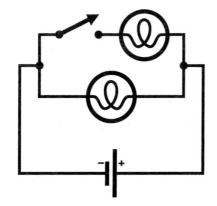

"I had them sign up for challenges. This got them looking at and talking about what was on the sheet and deciding together. Each time they took on another challenge they signed on the board again. This seemed to keep them motivated."
—Teena Staller, trial-test teacher

Challenge C: Turning both lights on/off in a series circuit

A switch anywhere in series with the pair of lights that are in series with each other will turn them both off.

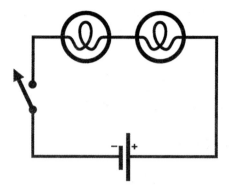

Challenge D: Turning one light on/off in a series circuit

A switch in parallel with one light that is in series with another light will turn just the one off.

This is a hard one. Students must see that the current will go through the closed switch, which has low resistance, instead of the lightbulb, which has many times more resistance. Note that the brightness of the light that remains lit changes when the switch is opened and closed. Note also that trying this method with a circuit that has only one light in it will make a short circuit.

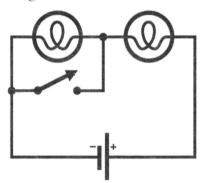

3. After the class discussion, ask teams to disassemble their circuits and put the materials into their "Experimenter's Kit" bags. Collect these, the batteries, and the journals.

■ Going Further

1. Journal Writing

Give your students time to write and draw in their journals about what they've learned, their discoveries, problems and solutions, ideas for future projects, and unanswered questions they still have. You may want to suggest that they write limericks or other types of poems or songs incorporating content covered in the unit. See also the next "Going Further."

2. Limericks and Songs

On page 159 you'll find some limericks *(A Series of Limericks That Parallel These Lessons)* and a song *(The Circuit Game)* which incorporate electric circuit concepts. These can either be read aloud to the students or used to inspire them to write their own artistic works on the topic.

■ Special Going Further

People Model of Electric Current

Series and Parallel Circuits

In the instructions that follow, we have tried to give a step-by-step description of what you and your students will do in this model. We have not, however, left openings for extensive class discussion or for questions from your students. You may choose to be flexible in the presentation of this model to match the needs of your students.

This model of how electric current flows at a subatomic level includes the following topics related to Sessions 7 through 9:

- resistance
- series circuits
- parallel circuits
- adding lightbulbs
- adding voltage to a series circuit
- short circuits
- switches in series and parallel

What You Need

❑ all the material prepared for the previous People Model of Electric Current at the end of Session 4 (page 98)
❑ 1 small box
❑ 2 tubes approximately 5 inches in diameter and 8 inches long

Getting Ready

1. Read the entire Special Going Further on pages 98–104 to refresh yourself on the materials used, the role of each student, the procedures followed, and the circuits made.

2. Gather the material used in the previous people model (pages 98–104).

3. In addition to the signs from the previous people model, make one more "battery" sign, two more "lightbulb" signs, and one new "switch" sign on strips of paper. These will be used to label volunteers in the model.

4. As in the previous people model, prepare two more tubes for the lightbulb students to pass or roll sponge pieces through. Label each tube "filament."

5. Label the small box "battery." Along with the two boxes from the previous people model, it will represent a third battery. Label one end of the box with a plus symbol and the other end with a minus symbol to represent the positive and negative terminals of a battery.

6. Have all the old and new material readily available to conduct the activity.

1. Model of a Series Circuit

Review and Setting Up a Circuit

1. Remind students of the people model activity they participated in earlier, and of the procedures they followed and roles they played. Tell them they'll do similar activities here, but will build different circuits.

2. Because, in this people model, you'll have several lightbulbs and batteries in the circuit, you'll need 12 to 15 volunteers. Either assign the same students to the roles of battery, lightbulb, and wire, or ask for new volunteers. Using the signs you made earlier, label three students as "lightbulb" and three as "battery." Label the remaining six to nine students as "wire" or "atoms in the wire." Set aside the "switch" sign for use at the end of the model.

3. Have one lightbulb student, two battery students, and all the wire students stand in a line at the front of the room. Ask the remaining two lightbulb students and one battery student to take a seat nearby while they await their turn to be added to the circuit.

4. Hand out the appropriate material to just the volunteers standing in the line. Give a tube (filament) and one sponge piece (electron) to the lightbulb student; one battery box (full of sponge-piece electrons) to each battery student; and one sponge piece (electron) to each wire student. If necessary, remind students they'll be portraying the electrons that move freely between and around the atoms in a conductor.

5. Make sure you have the four "brightness" signs available. Show each one to the lightbulb student (and the rest of the class) and remind him of his job to call out which one should be displayed.

6. Ask the class how the volunteers should be arranged to make a closed circuit with one lightbulb and two batteries. The volunteers should end up arranged in a circle with wires connecting the light-bulb and the batteries.

7. Ask students to notice the orientation of the battery boxes. If the terminals of both batteries are not pointed in the same direction, ask students if they need to be. [Yes, the positive and negative ends of both batteries should point the same way.]

8. Remind the volunteers that when the current is flowing they should give an electron away with one hand while receiving one with the other hand, and to **always make sure they have one in one of their hands.** This means they'll all pass electrons at the same time.

9. Remind them, too, that the battery students will begin the flow by getting an electron out of their boxes and passing them to the wire students at the negative end. The electrons will then flow around the circuit. The wire students at the positive ends of the batteries will drop the electrons back into the battery boxes.

10. Have the volunteers begin. Once the circuit gets going, ask the lightbulb student which sign you should display. Hold up the Very Bright sign above the lightbulb student.

Making a Series Circuit and Adding More Lightbulbs

1. Now pause the model by making it an open circuit. Announce that you're an item made of rubber (an insulator), and step into the circle between two wire students to open the circuit. Ask the students what they think will happen if you add another lightbulb to the model.

2. After hearing their predictions, ask one of the seated lightbulb volunteers to stand up. Give her a filament and an electron. Ask the class where the second lightbulb student can join the circuit. Verify that she can be anywhere, as long as she's between two wires. Have the new lightbulb join in the circle. Take yourself out of the circle and join the wires to make a closed circuit again. Have the students begin passing electrons. When the current is flowing well ask the two lightbulb students to discuss and decide what sign you should display. Hold up the Bright sign.

3. Step in the circle to open the circuit again, and ask the students to predict what will happen if a third lightbulb is added. Ask the remaining lightbulb volunteer to stand up and give him a filament and an electron. Let him decide where he should go in the circuit. When the third lightbulb is correctly in place between two wires, step out of the circle, close the circuit, and have the students begin passing the electrons again. When the current is flowing well ask all three lightbulb students to discuss and decide what sign should be displayed. Hold up the Dim sign.

4. As the volunteers continue to pass electrons, ask the class what kind of circuit this is. [A series circuit.] Review why it's called that. [Because the electrons flow through one bulb after another.]

Resistance

1. Step into the circle to make an open circuit and stop the flow of electrons. Draw students' attention to the lightbulbs and come to agreement that it's harder for the electrons to get through the filament. Ask them what this is called. [Resistance.]

2. Tell students this helps to reinforce the concept that when more resistance is added to a series circuit, the current decreases. Remind them that when the current decreases, the battery is drained more slowly, and lasts longer.

Removing One Bulb and Adding Voltage

1. Ask one of the lightbulb students to step out of the circuit and have a seat nearby. Ask the class if the circuit is still a series circuit. [Yes, now it has two lightbulbs in series.]

2. Now take one of the remaining two lightbulb students completely out of the circuit and close the wires where he had been. Ask the class what they think will happen when the flow of current resumes. Step out of the circle, close the circuit, and ask students to begin passing electrons. When the current is flowing well ask the remaining lightbulb student which sign she thinks you should display. Hold up the Very Bright sign. Remind students that this is just like the first circuit they set up in this people model.

3. Step back into the circle to make an open circuit. Have the second lightbulb join the circuit again. Ask the class to predict what will happen to the second lightbulb if the other bulb is pulled out of—but not completely removed from—the circuit. After listening to their predictions, ask one of the lightbulb students to step back from the circle. Let the wire students who are on both sides of that lightbulb know that they are **not** to move together to close the gap. Tell the "pulled" lightbulb that when the current flows again, he is to *refuse to take electrons* from the wire student who is passing them to him.

4. Take yourself out of the circle, join together the wires you were between, and ask the students to resume the flow of electrons. [The result should be that since the "pulled" lightbulb refuses to take electrons from the wire and since that wire can't pass any electrons she can't receive them either, the flow of electrons will stop. Therefore the second lightbulb can't shine.] Ask the second lightbulb what sign you should display. Hold up the Off sign. Tell students this helps to reinforce the concept that when one bulb is removed from a series circuit, any other bulb(s) in the circuit goes out.

5. Step into the circle again to open the circuit. Put the other two lightbulbs back into the circuit for a total of three. Show the class the third battery box and ask what they think will happen when it's added to the circuit. After hearing their predictions, ask the other battery volunteer to stand up. Give her a battery box (full of electrons), and ask her where she should go in the circle. Also ask her in what direction the terminals of her battery should point. Discuss the direction if your class needs reinforcement of the concept.

6. When the third battery is correctly in place, step out of the circle and ask students to begin the flow of electrons. When the current is flowing well ask the three lightbulb students to discuss and come to an agreement about what sign you should display. Hold up the Bright sign.

7. As they continue to pass electrons, remind the students that in an earlier circuit when all three lightbulbs were in series and the circuit had only two batteries, the three lightbulbs were dim. Let students know that this helps to reinforce the concept that when more voltage is added to a series circuit, the current increases (and drains the battery more quickly).

Short Circuit

1. Step into the circle to open the circuit, then ask the students how they could make a model with almost no resistance—a short circuit. Listen to and validate their ideas.

2. Step out of the circle and ask students to begin passing electrons. Remove all three lightbulbs from the model and keep the electrons moving through the circuit. Ask the students to notice what's happening with the electrons. [They're moving much faster without the resistance of the lightbulbs.] Ask what effect this could have on the

circuit. [It's a short circuit; it could run down the battery more quickly, heat up the wire, maybe cause a fire.]

2. Model of a Parallel Circuit

Making a Parallel Circuit and Removing One Bulb

1. Once again step into the circle to open the circuit and stop the flow of electrons. Ask a lightbulb and battery student to step out of the circle and take a seat nearby so that there will be just two lightbulbs and two batteries in the new circuit that's going to be made. Ask the class how the volunteers should be arranged to form a circuit with a branch where the current can go two different ways. Once the students are arranged correctly, ask what this kind of circuit is called. [A parallel circuit.]

2. Draw the students' attention to the place where the circuit branches into two paths. Tell the volunteer whose place in the circuit is just before that branch that she will alternately pass sponge-piece electrons to one path and then the other. Explain further that because the volunteers can't pass an electron until they are given one, that one of the paths *will* be passing electrons while the other path isn't. So the two paths will take turns either passing electrons or not passing them.

3. Point out to students that this does **not** mean that the current is not flowing through one of the paths! Let them know that it just shows the limitation of the model. Explain that since, in the model, students are passing just one electron at a time, the current *appears* to stop in one path of the circuit. Be sure that students understand that in a real circuit (with real wires, lights, and batteries) MANY electrons are flowing and they **all** go through **both** paths.

4. Before closing the circuit and allowing the current to flow, ask the students to predict what will happen. After hearing their predictions, close the circuit and resume the flow of electrons. When the volunteers get the hang of the new way to pass electrons and the current is flowing well, ask the two lightbulb students to decide what sign you should display. Hold up the Very Bright sign.

5. As students continue to pass electrons, ask why these bulbs in a parallel circuit are both burning very bright, when in the series circuit, the addition of a second bulb made both lights shine merely bright. Remind students that each lightbulb provides resistance. In the parallel circuit, the current can go two ways. Either way the current goes, it flows though only *one* lightbulb and therefore

encounters less resistance. In the series circuit, however, the current flows through *both* lightbulbs and encounters greater resistance.

6. Step into the circle to open the circuit, then ask the students to predict what will happen if one of the two lightbulbs is pulled out. After hearing their predictions, ask one of the lightbulb students to take a step back. Tell the wire students on either side of that light **not** to close the gap. Take yourself out of the circle, join the wires where you had been, and ask students to begin passing electrons, again reminding the volunteer just before the branch to alternately pass to one of the paths and then the other. When the current is flowing well ask the remaining lightbulb student what sign you should display. Again, hold up the Very Bright sign. Ask the class why it's still shining. [Because it's still part of a closed circuit.] Tell students this helps to reinforce the concept that when one bulb is removed from a parallel circuit, any other bulb(s) in the circuit still shine.

Parallel Circuit Switch That Turns Off Both Lights

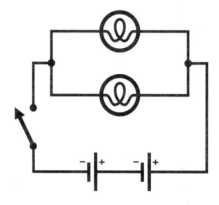

1. Step into the circle to open the circuit. Ask the "pulled" lightbulb student to step back into the circuit. Ask the class to remember how the lights shined in this kind of circuit. [Both were very bright.]

2. Ask the students how they could turn off both lightbulbs at once. [Use a switch to open the circuit at any point where there's one path.] Ask a few students for different locations they can suggest that would work, then decide on one of the locations. Using the switch sign you made earlier, label a wire student at that location as a switch as well.

3. Take yourself out of the circuit to close it, and have students begin the flow of electrons. When the current is flowing well tell the switch student to step back from the circuit (perhaps you could say something like "flip the switch"). Ask the lightbulb students what sign you should display. Hold up the Off sign.

Parallel Circuit Switch That Turns Off One Light

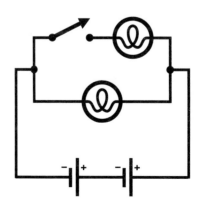

1. Step into the circle to open the circuit. Now ask students how they could turn off only one lightbulb. [Use a switch to open the circuit at any point where it is split into two paths.] Again, ask your students' input about where to do this on the circuit. When they've decided on one or more locations, label a wire student at one of the locations as a switch as well.

2. Take yourself out of the circle to close the circuit and have students begin the flow of electrons. When the current is flowing well tell the new switch student to step back from the circuit. Ask the lightbulb students what sign you should display. This time the answer is different for the two of them. For the lightbulb student who is **not** in series with the switch, hold up the Very Bright sign. Hold up the Off sign for the other lightbulb student (the one *in series* with the open switch).

3. Back to a Series Circuit

Series Circuit Switch That Turns Off Both Lights

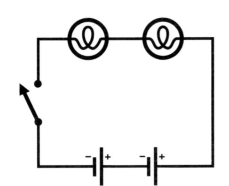

1. Step into the circle to open the circuit. Ask the volunteers to again make a series circuit with two lightbulbs and two batteries. Remind them that as before in a circuit like this both lightbulbs would be bright.

2. Ask the students where they could put a switch that would turn off both lights. [At any point in the circuit.] When students decide where, label a wire student at that location as a switch as well.

3. Take yourself out of the circle to close the circuit and have students begin the flow of electrons. When the current is flowing well tell the switch student to step out of the circuit. Ask the lightbulb students to decide what sign you should display. Hold up the Off sign.

Series Circuit Switch That Turns Off One Light

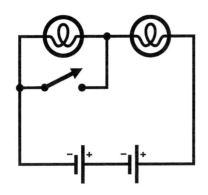

1. Now it's time for a hard one. Once again step into the circle to open the circuit. Ask students where they could put a switch that would only turn off one light in a series circuit. [Make a switch in parallel with one bulb.] When the students decide where, label a wire student at that location as a switch as well.

2. For this circuit you'll do the next couple of steps in a different order than for the previous three circuits. First ask the switch student to step back from the circle to represent an *open* switch. Take yourself out of the circle to close the circuit and have students begin the flow of electrons. When the current is flowing well ask the lightbulb students which sign you should display. Hold up the Bright sign. Help students to see that when the switch is open, this is a "normal" series circuit.

3. With the current still flowing well, ask the switch student to step back into the circuit to represent a closed switch. Again, ask the lightbulb students what sign to display. Now the answer is different

for the two of them. For the lightbulb student in parallel with the switch, hold up the Off sign. Hold up the Very Bright sign for the other lightbulb student (the one in series with the switch and light-bulb that are in parallel). Remind students that when the switch is closed, the current flows *through it* because it has less resistance than the bulb. Also, the bulb in parallel with the switch goes off because it receives no current, and the other bulb gets brighter because it "gets all" of the current.

4. Collect the materials volunteers have been holding and the signs that identified their roles. Thank all the volunteers for their good work and dismiss them back to their desks.

Switches in Series and Parallel Circuits

Challenge A
Lights in Parallel

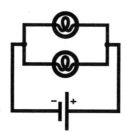

Redraw the diagram with a switch in it that will turn **both lights** on and off. Build the circuit to test your idea.

Challenge B
Lights in Parallel

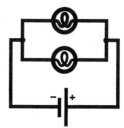

Redraw the diagram with a switch in it that will turn **just one light** on and off. Build the circuit to test your idea.

Challenge C
Lights in Series

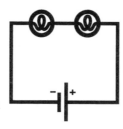

Redraw the diagram with a switch in it that will turn **both lights** on and off. Build the circuit to test your idea.

Challenge D
Lights in Series

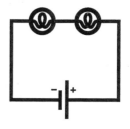

Redraw the diagram with a switch in it that will turn **just one light** on and off. Build the circuit to test your idea.

18

A Series of Limericks That Parallel These Lessons
by John Erickson

While studying electricity
I asked the teacher to tell me,
        "When you spell parallel
        Is it two or one *l*?"
The teacher said, "No. There are three.

"It starts *p – a – r – a* – double *l*,
The two *l*s side by side, parallel.
        Then ends with *e – l*,
        Which together spell *el*,
Making three *l*s in pa – ra – l – l – el."

Still confused, with yet unanswered queries,
I turned to electric theories.
        My problems dissolved
        When I firmly resolved
To make all future circuits in series.

But that did not work out too well.
I was caught in a difficult spell.
        For my project to work, it
        Required that my circuit
Have connections in parallel.

In series I wanted the current
To make the lights bright, but they weren't.
        The resistance, combined,
        Of the lightbulbs consigned
Them to dimness, or so I have learnt.

But in parallel circuits, however,
The current can flow to wherever.
        The resistance is low
        So the current can flow,
But the battery won't last forever.

So I hooked it up in parallel,
With connections in series as well.
        Then I threw in some switches
        And produced something which is
Considered a circuit that's swell.

The Circuit Game
(apologies to Joni Mitchell)
by Kevin Beals

And the currents they go round
and round
And the battery voltage goes up
and down
We're captive on a carousel of wire
We can't return and we never tire, but
go back to where we came
And go round and round and round
in the circuit game.

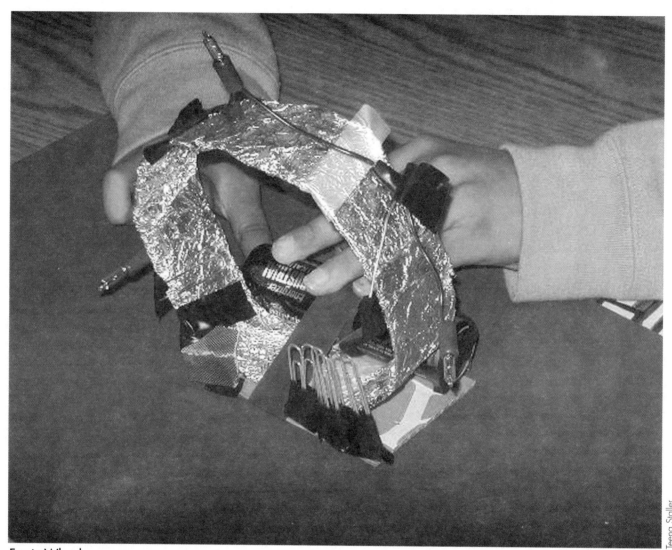

Ferris Wheel

SESSION 10: ELECTRIC INVENTIONS

If someone says it can't be done, take it as a challenge, not as discouragement. It's the only way to make inventions. —Patsy Sherman, co-inventor of Scotchgard™

Overview

This session gives students the chance to do what many of them have probably been craving to do throughout the entire unit: design and assemble their own "invented" circuit. The session begins with a class brainstorm of potential projects, and then in teams of two the students begin their own project. There's a lot going on during this session, and plenty of excitement, so some teachers find adult volunteers very helpful.

As with the switch-inventing in Session 5, some students will have an easier time than others with the creativity and patience, but all will benefit from the experience. To allow for this spectrum of ability and personality, student pairs are encouraged to invent something totally new, copy a device they've seen modeled, *or* do something in between the two. Ideally, they'll "invent" something by at least slightly modifying a copied device.

The students work at the level of complexity of their choosing (given the available materials) to plan and build any invention with an electric circuit in it. Some may include parallel and/or series circuits with fancy switches, but many devices will consist of a simple circuit. With even a simple circuit, however, there's much that the students can demonstrate about where the electricity flows and how it helps the invention work.

> Students may come up with project ideas that require materials not included on the provided list. For this reason, some teachers prefer to do a short session in which students come up with ideas and begin constructing, then let students go home and gather more materials. Students can either finish their projects in class or as homework.

■ What You Need

For the class:

- ☐ the **No 120V, Battery, Lightbulb, Wire or Other Conductor, No Short Circuits!, Switch,** and **Connection/No Connection** posters from previous sessions
- ☐ the homemade electric devices used in Sessions 2 and 3
- ☐ 1 roll of aluminum foil
- ☐ an assortment of metal paper clips, large and small
- ☐ several paper fasteners (brads)
- ☐ 30 or more holiday lightbulbs with stripped leads
- ☐ 30 or more wire pieces with stripped ends
- ☐ extra wire to cut and strip as needed
- ☐ several pairs of wire cutters and strippers
- ☐ several scraps of cardboard, of various thicknesses

□ various items for building (popsicle sticks, pipe cleaners, straws, etc.)
□ clear tape and masking tape
□ *(optional)* simple motors and buzzers

For each pair of students:
□ 2 C-cell batteries
□ 2 prepared film canisters and lids
□ 1 pair of scissors
□ masking tape or stickers

For each student:
□ the **Electrical Inventor's Journal** from the previous session
□ 1 copy of the **Electric Invention Assignment Sheet** (page 166)
□ 1 copy of the **Invention Spec Sheet** (pages 167–168) to add to journal

■ Getting Ready

1. Make one copy each of the **Electric Invention Assignment Sheet** (page 166) and the **Invention Spec Sheet** (pages 167–168) for each student.

2. Set out the **Electrical Inventor's Journals** from previous sessions.

3. Gather the homemade electric devices you used in Sessions 2 and 3. Set them out where students can refer to them for ideas. Also gather any remaining pieces of wire students stripped at Station 6 in Sessions 2 and 3.

4. Select a part of the room to be the supply area. Stock it with cardboard pieces, extra paper clips and fasteners, foil, holiday lightbulbs, wire pieces, and other building materials.

5. Assemble the batteries, prepared film canisters, scissors, and tape or stickers for each pair of students.

6. Be sure all the posters put up in previous sessions are still prominently displayed.

(GO) ■ Inventing Electric Devices

1. Pass out an **Electric Invention Assignment Sheet** to each student. Lead a class brainstorm of possible projects that could be done with electric circuits. Record their ideas on the board.

2. Remind the class that although some electric devices (the flashlight, for example) are fairly simple, and were invented many years ago, the students can still invent their own versions of a simple device. (A "newer and cooler" flashlight....)

3. If students have trouble coming up with ideas, you may choose to share some ideas used by students in other classes:

- a purse that, when opened, turns a light on inside
- an aluminum foil "monster," with eyes that light up
- a doll house lamp
- a "space ship" that lights up
- board games
- quiz boards
- a see-saw switch
- a mini garage with a light that goes on when a toy car enters
- a lighthouse with a flashing or rotating light
- a face with a nose that lights up
- a basketball hoop with a light that goes on when a small aluminum ball is tossed in

GEMS would love to hear about inventions your students come up with!

4. Remind students they can also get ideas from the notes they may have made in their journals when they traveled through the trade show in Session 6, or at other times throughout the unit.

5. Pass out the **Electrical Inventor's Journals** from the previous sessions. Also pass out a copy of the **Invention Spec Sheet** to each student and ask them to add it to their journals. Remind students that as before, although they will be working with a partner on an invention, each student will fill out her own spec sheet.

6. Emphasize that an important part of the project is their description of the *circuit* in their device. Remind students of the spec sheet they filled out earlier for their switch invention (pages 14 and 15 of their journals). If necessary, review what a spec sheet is. Go through the **Invention Spec Sheet** with the students to make sure they know what information they need to record. Be sure to point out, on the second page, that they must include a schematic diagram and write about how the current flows in the circuit.

■ Constructing Their Inventions

1. Organize the students into pairs. Distribute the batteries, film canisters, scissors, and tape or stickers to each pair. Then show students the building materials in the supply area.

2. Remind students to watch out for short circuits and tell them they're limited to two batteries in their inventions.

3. Let the student pairs spend the first few minutes making plans. Ask if any pair has an idea they want to tell the class about. Send pairs who can't think of an idea to look at the homemade electric devices you set out earlier. If you think students might be intimidated by complex ideas, tell them it's alright if they want to make something that's a lot like these devices.

4. Have the pairs begin working on their projects. Circulate and assist as necessary, and cut and strip wire as needed.

■ Cleaning Up

1. Twenty minutes before the end of class, warn the students that you'll be stopping them soon. Let them know they'll be able to finish up their projects in the next session (if you decide to add such a whole or partial session) or that they'll take the projects home to complete.

2. Fifteen minutes before the end, ask students to stop working and label their projects and any spare parts with their names.

3. Have pairs turn in parts they don't plan to use, and throw out or reuse any scraps of material.

4. Ask students to remove the batteries from their inventions. Collect the batteries.

5. Have students spend any remaining time working on their spec sheets.

■ Going Further

1. Award-Winning Electric Inventions by Kids
If you think these will inspire, and not intimidate, your students, read some of the following descriptions of elaborate, award-winning inventions by kids.

Turn-Signal Turner-Offer, invented by Laurie DiStefano of Delaware in the twelfth grade. "My father would always turn his turn signal on when he was changing lanes, and we'd end up driv-

If you have capable students who need motivation, and you think they will be inspired and not intimidated by hearing about more complicated projects, you may want to read them the stories in "Going Further" below. These are stories about award-winning electric inventions by other students, and your students may choose to try to make these themselves.

ing with it on for hours." Laurie invented a device that automatically switches the turn signal off after two minutes.

Glo Panel, based on the **Glo Sheet** invented by Rebecca Schroeder when she was ten years old. As Rebecca did her math homework in the car while she waited for her mother, it got dark, and Rebecca had to stop. Later, at home, she painted phosphorescent paint on a piece of cardboard, let it dry, then was able to put paper on top of it and work on her "glowing" paper in the dark. Her invention is now used by nurses in hospitals, to check on patients without turning on the lights, and by the United States Navy when personnel work at night on the decks of ships. She later developed a battery-operated version called the Glo Panel, and now owns her own company.

Safety Seat Belt, invented by Brandy Curry and Jason Lamontagne of Pennsylvania when they were in the ninth grade. Their device sounds an alarm when a child is not seat-belted.

Whirling Dart Board, invented by Patty Brandetsas of North Carolina while in eighth grade. She took an ordinary dartboard and rigged it to a battery device that rotated the dartboard while players threw darts.

Porch Policeman, invented by Emily Strubinger of Ohio in the second grade. After someone stole the pumpkins off her porch, Emily designed a button alarm to place under a pumpkin; if someone lifted the pumpkin, the alarm would sound.

Child Safety Door Alarm, invented by Whitney Blake of New York while in the ninth grade. Blake says she invented the alarm so her young cousins wouldn't fall into the family swimming pool. "It also works on a gate or a door, and it can be left on at all times because it doesn't sound when a taller person passes," says Blake.

Cast Cooler, invented by Krysta Morlan in the ninth grade. Because Krysta had surgery for diplegic cerebral palsy (a condition that results from injury to the brain), she had to wear a full leg cast for almost a year. She invented the Cast Cooler because the cast often made her leg hot and uncomfortable. Powered by a 9-volt battery, an electric motor pushes air through a tube and into the cast.

2. Journal Writing

Give your students time to write and draw in their journals about what they've learned, their discoveries, problems and solutions, ideas for future projects, and unanswered questions they still have.

Electric Invention Assignment Sheet

Your project must have these things:

1. An invention that has a circuit in it.

Here are some examples:

- A simple circuit with a new switch design.

- A more complicated circuit with switches, lights, and batteries.

- A game or challenge with a circuit in it.

- A simple model with a circuit in it, such as a car, house, or robot.

2. A title and a list of the materials used—both the conductors and insulators.

3. A drawing of your invention.

4. A schematic diagram of your invention.

5. A description of how your invention works.

6. Anything else you can say about the circuit:

- What materials carry the current?

- Are there series parts or parallel parts in the circuit?

- Where does the current flow?

- How brightly does the light shine?

- How do the switches change the flow of current?

Electrical Inventor's Journal

ideas for inventions • drawings • questions • experiments • problems • solutions • stories

Invention Spec Sheet

company logo

Name of Invention: _____

Inventors: _____

_____

Date: _____

Materials

Conductors	Insulators
_____	_____
_____	_____
_____	_____
_____	_____
_____	_____
_____	_____
_____	_____

Draw your invention. Use a different-colored pen or pencil to show where the current flows.

19

Electrical Inventor's Journal

ideas for inventions • drawings • questions • experiments • problems • solutions • stories

Draw a schematic diagram of your invention.

How does it work? Where does the current flow through the circuit? Describe the series and/or parallel circuits in your invention. Describe the switch and how it affects the flow of the current. Describe the role of other items, like bulbs, batteries, etc. in the circuit.

_____

_____

_____

_____

_____

_____

_____

_____

_____

20

Teena Staller

If I find 10,000 ways something won't work, I haven't failed. I am not discouraged, because every wrong attempt discarded is another step forward. Just because something doesn't do what you planned it to do doesn't mean it's useless. —Thomas Edison

Overview

In this session the students finish the projects they began in Session 10, then bask in the glory of the variety of inventions they have built.

The session begins with a few minutes for pairs to finish their spec sheets and finish their projects. Each pair is given an opportunity to describe their invention and demonstrate how it works.

The pairs are then assigned the roles of "Explainers" and "Rovers," as in Session 6. The Explainers stay by their inventions to answer questions about them, demonstrate how they work, and make on-the-spot repairs. The Rovers try to find out as much as they can about their classmates' inventions—especially how the circuits work. Halfway through the process, they switch roles.

■ What You Need

For the class:
- ❏ the **No 120V, Battery, Lightbulb, Wire or Other Conductor, No Short Circuits!, Switch,** and **Connection/No Connection** posters from previous sessions
- ❏ the extra material (aluminum foil, metal paper clips, paper fasteners, extra wires, etc.) from Session 10

For each pair of students:
- ❏ 2 C-cell batteries
- ❏ their invention from Session 10, ready to be finished
- ❏ *(optional)* 1 piece of paper labeled as "Rover's Recording Sheet"

For each student:
- ❏ the **Electrical Inventor's Journal** from the previous session

■ Getting Ready

1. Set out the batteries, inventions, and the extra materials from Session 10 that students will need to put the finishing touches on their projects.

2. Set out the **Electrical Inventor's Journals** from previous sessions.

3. If you'd like students to use it, write "Rover's Recording Sheet" at the top of a piece of paper and copy enough for each pair of students.

■ Finishing the Inventions

1. Tell students that after they've had some time to finish their inventions, they're going to describe and demonstrate them for the class.

2. Pass out the **Electrical Inventor's Journals** and the students' inventions. Give student pairs five minutes to complete the spec sheets for their inventions and discuss what remains to be done.

3. Pass out the C-cell batteries and remind students about the supply area. Then give them about ten more minutes to finish their inventions.

4. Have students return unused materials and tidy up.

5. Call on each pair to briefly describe their invention and demonstrate how it works for the class.

■ Invention Festival

1. Let students know that now they'll participate in an Invention Festival, which will be similar to the Electric Switch Trade Show in Session 6. As in the trade show, remind students they should leave their spec sheets with their inventions. If you've decided to use them, pass out a "Rover's Recording Sheet" to each pair of students. Divide each pair into a Rover and an Explainer. Remind students they'll switch roles halfway through the activity.

2. Review that the Rovers will view their classmates' inventions and see how the circuits in them work.

3. Review also that the Explainers will stay by their inventions to answer questions about them and demonstrate how they work, and will make any on-the-spot repairs. Point out that it will be the Explainer's choice whether to let visiting Rovers operate their invention. The Explainer may choose to simply demonstrate how it works.

4. Let Rovers rove for about 10 minutes, then have partners change roles and go about 10 minutes more.

■ Going Further

1. Showing the Parents

As a follow-up to the unit, many teachers like to organize an opportunity for students to invite their parents to the classroom to see their work.

2. Journal Writing

Give your students time to write and draw in their journals about what they've learned, their discoveries, problems and solutions, ideas for future projects, and unanswered questions they still have.

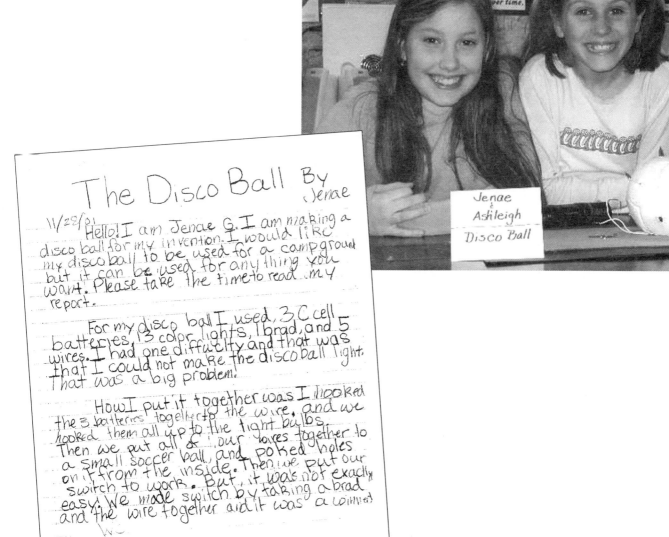

Jenae
&
Ashleigh
Disco Ball

The Disco Ball By Jenae

11/28/01   Hello! I am Jenae G. I am making a disco ball for my invention. I would like my disco ball to be used for a campground but it can be used for anything you want. Please take the time to read my report.

For my disco ball I used, 3 C cell batteries, 13 color lights, 1 brad, and 5 wires. I had one difficulty and that was that I could not make the disco ball light. that was a big problem!

How I put it together was I hooked the 3 batteries together to the wire, and we hooked them all up to the light bulbs. Then we put all of our boxes together to a small soccer ball, and poked holes on it from the inside. Then we put our switch to work. But it was not exactly easy. We made switch by taking a brad and the wire together and it was a winner!

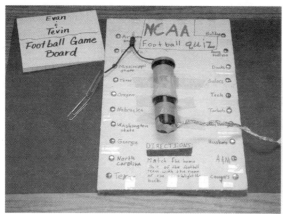

Evan & Tevin
Football Game Board

NCAA Football Quiz

Kirk & Wess
The Beetle Bug

Jenae & Ashleigh
Disco Ball

Mallory & Debbie
Night Light Robot

Kade and Brock
POLICE CAR

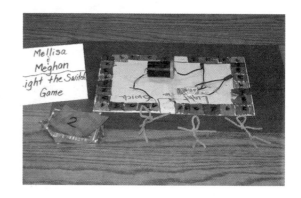

Mellisa & Meghan
Light the Switch Game

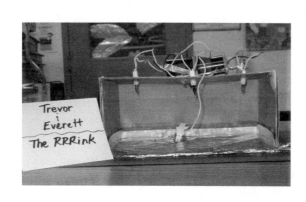

Trevor & Everett
The RRRink

All photos by Teena Staller

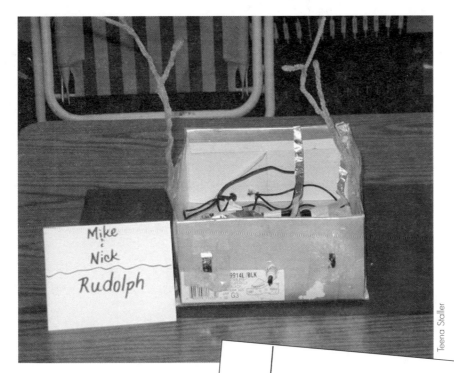

Teena Staller

Mike

The name of my partners and I's invention
is called Rudolf. It is a box with two
clear lightbubs for eyes and 1 red one
for the nose. We named it rudof because
of his red nose.

We have 3 switchs. 2 for the eyes
and 1 fore the nose. Both of the eyes
are in a paurclaleircnt. The nose is in a
series. We used three c-cell batteries to
give enough juice to light up all
the lightbulbs. On the switchs for the eyes
the right swich has to be on for the
left switch to work.

My favorite 3 inventions made in our class
was Tevins and Evens, Trevors and Everitts
and arés. I think that they all are tied thow. Trevors and
Everitts are cool because they have lights hanging
from the ceiling of there box and it reflects
of the tinfold. Evens and Tevens because there
is a NCCA football quiz board to get the
lightbulb to light you have to match the
state and the nickname.

In the future I would like to make an
invention with 3 light bulbs and be able
to make a switch that will turn it in to a
parrchl circut and a sries cirut. Then I would

What Is Electricity?

elec-tric-i-ty *noun* ... **3:** keen contagious excitement
—*Merriam-Webster's Collegiate® Dictionary*

Electricity is defined in several ways. Sometimes it's used to mean electric charge, and sometimes to mean electric current. Maybe the best definition of electricity is that it's a branch of science that deals with electric charge and electric current, their properties, and their uses.

Glossary for *Electric Circuits*

Essential vocabulary

Circuit—The path an electric current follows.

Closed circuit—A circuit that allows electric current to flow from a battery (or other power supply) around a complete unbroken conducting path and then back to the battery.

Conductor—An object or material through which electric current can flow with little or no resistance. (Metals and liquids with dissolved ions are the most common conductors.)

Current—The flow of electric charge. In metal conductors, the charge is carried by electrons.

Insulator—An object or material through which electric current cannot flow.

Open circuit—A circuit that has no complete path for electric current to flow around.

Parallel circuit—A circuit in which the current splits and runs through separate components.

Resistance—The tendency for an object or material to reduce the flow of current in a circuit. An electric current gives off energy (usually heat and light) when it flows through a material with resistance.

Series circuit—A circuit in which the current flows through each component (such as a lightbulb) one after the other.

The term "electricity" is not scientifically well defined. It's more proper to say that there's an "electric current" than to say that "there's electricity" or that the "electricity is flowing." Although it's a good habit to use the words correctly, most people, including scientists, use the word "electricity" loosely, and it's generally not worthwhile to correct students who do so.

Short circuit—A simple closed circuit with little or no resistance.

Switch—A device that either opens or closes a circuit.

Volt (voltage)—Voltage is a useful way to describe how strong an electric power supply is. In the United States, an outlet is about 120 volts. A regular flashlight battery is about 1.5 volts. (There is a more scientific definition, which elementary and middle school students should not be required to know: Voltage is a measure of the energy used for each unit of charge that moves through an electric circuit.)

Useful additional vocabulary

Amp or ampere—A measure of electric current. A standard 60-watt lightbulb runs on a current of about one-half amp.

Atom—A particle with protons and neutrons at the center (the nucleus), surrounded by electrons. Atoms make up all the elements, and elements make up all common forms of matter.

Battery—Electric cells connected in series. The word battery is also popularly used to refer to a single electric cell.

Cell—A device that creates a steady voltage and provides an electric current. Most electric cells work by means of a chemical reaction, but there are cells powered by light (solar cells) and cells that use fuel (fuel cells).

Charge—A fundamental property of matter. *The whole topic of electricity is about forces and movements of electric charge.*

Circuit breaker—A switch that is normally closed, but which opens when the current reaches a certain level. Circuit breakers are put in circuits to make sure the current stays at a safe level and overheating doesn't occur. (See "Fuse.")

Electron—A small particle that carries an electric charge.

Fuse—An in-circuit device deliberately constructed to melt when the current reaches a certain level. This opens the circuit so that no more current can flow. Like circuit breakers, fuses are put in circuits to make sure the current stays at a safe level and overheating doesn't occur. They differ from circuit breakers in that when they "blow," they must be replaced, instead of simply being turned back on.

Ion—An atom (or group of atoms) that carries an electric charge because the number of its electrons is different from the number of protons.

Semiconductor—A material that has some resistance.

Watt—A unit of power. Wattage measures how quickly electric energy is being used.

Electric Charge

Electric charge is one of the fundamental properties of matter. There are two opposite types of electric charge—positive and negative. All matter is a combination of particles with either positive or negative electric charge. We know an object has electric charge when we can see that there is a force on the object from another electric charge. (For example, you know a sock from the dryer is charged when it's attracted to a sweater.) The rules for these forces are well known—opposite charges are attracted to each other; charges that are alike are repelled by each other.

Most often, there is an equal amount of positive and negative charge in an object or substance, so the total electric charge adds up to zero. That's why electric forces are not a noticeable part of every moment's experience, the way the force of gravity is.

At the atomic level of matter there are three main particles: protons, which carry a positive charge; neutrons, which have no overall charge; and electrons, which carry a negative charge. The charge of a proton is exactly equal and opposite to the charge of an electron—so a proton and an electron together have no overall charge. The central mass, or nucleus, of an atom is made of protons and neutrons. The positively charged protons in an atom's nucleus tend to attract an equal number of negatively charged electrons, so a collection of atoms is usually electrically neutral.

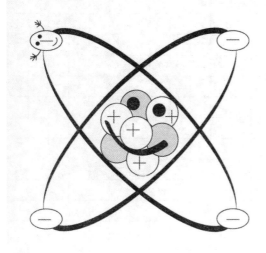

"Classic Style" Atom

It shows the balance of positive and negative charge, but the size, shape, and behavior of its parts is misleading.

More accurate, but conceptually difficult electron cloud atom.

Static Electricity

Some materials have a tendency to transfer electrons to certain other materials. Hair, for instance, will give some of its electrons to a balloon. Rubbing the materials together increases their contact and facilitates the transfer of electrons. The result is more electrons on the balloon, creating an imbalance of electric charge, which means that the balloon has more negative charge than positive. The electric forces can be observed when the balloon attracts the hair or clings to the wall as the electrons from the balloon cling to atoms in the hair or wall. This is static electricity.

You may have observed the generation of static electricity when shoes rub on a carpet, clothes come from the dryer, when someone pets a cat, or when a child goes down a plastic slide on a playground.

The word "electricity" comes from the ancient Greek word for amber. Amber, a form of petrified tree sap, was one of the first materials known to exhibit static cling. For millennia the only effects that people associated with electricity were static electric effects. When Benjamin Franklin established that lightning was an electrical effect, and that the pushes and pulls of static electricity were due to two opposite types of electric charges, the science of electricity took off.

Electric Current

An electric current is the movement of electric charge, either positive or negative. It doesn't count as electric current if positive and negative charges are moving together. An ordinary current of water, for example, has positive and negative charges moving together, and it is not an electric current. (See Electricity Is…, below.)

Current is the measure of both the speed and the quantity of electrons at any given point. In the guide, when we refer to "more" current we also mean *faster* current; "less" current means *slower* current.

Misconceptions about Electricity

The fundamental nature of matter is not, and may never be, fully understood. One thing is certain: This is a strange universe, and it's easy to lead people astray when you try to describe it accurately.

Most of the misconceptions described here are fairly benign. Students can harbor these misconceptions and still successfully complete and learn from the activities in this guide. The first one, however, the "Source and Sink" misconception, needs to be dispelled within the first few sessions if students are to grasp the basic lessons of building circuits.

Source and Sink
As mentioned above, this is an important misconception to address as early as possible in the unit if students are to successfully understand basic concepts about building circuits.

> *Misconception:* "Electricity comes from the battery [the source] and is used up by the lightbulb [the sink.]"

This common misconception is a true obstacle to successfully building electric circuits. If students consider the battery as the source of the electricity, it's natural for them to assume that one wire connecting the battery to the lightbulb will make the light shine. This illustrates why it's better to use precise terms such as "electric current" rather than the vague term "electricity." Students will understand, after repeated experiences, that current flows *through* the bulb and is **not** *used up by* the bulb.

Many people are familiar with screw-base lightbulbs. With these bulbs it's not obvious that there are two connections from bulb to wire—one on the side of the base of the bulb, and one on the very bottom. This may be one reason the misconception persists, even among people who have no trouble changing a lightbulb. The use of bulbs from holiday light strings is an advantage because each bulb has two distinct wires coming from it.

Electricity Is...

Students are sometimes given a marginal definition of electricity for the sake of a quick answer. Usually these definitions neither help nor hinder students who are trying to build circuits. One problem with these quick definitions, though, is that they incompletely describe the richness of electrical phenomena. Sometimes the definition is misinterpreted and draws students to conclusions that they have to "un-learn" if they go on to study the topic at a higher level.

Problematic definition: "Electricity is flowing electrons."

This is one of the most common definitions and a good example of some of the problems mentioned above. In its favor, it *is* a description of one common electrical phenomenon, and when students visualize flowing electrons it might help them to correctly assemble a circuit. On the other hand, there are flowing electrons that are not electric currents and there are electric currents that are not flowing electrons.

Flowing electrons that are not electric current:
The flow of water in even a small river carries more electrons than the current in the largest electric power lines. The current in a river is not an electric current, because the positive and negative charges are flowing along together.

Electric current that is not flowing electrons:
But an electric current does flow through water. It's not electrons, but charged atoms called ***ions.*** In computer chips and other electronic devices there are materials in which positive charges are conducted instead of electrons. The focus on electrons alone as carriers of electric charge leaves out a great deal of the picture.

Current in Water

Misconception: "Electrons flow through water."

In fact, the electrons in a liquid like water are *not* free to flow. As mentioned above, charged atoms (or sometimes groups of atoms) called ions carry electric current through water. Ions can have either a positive or negative charge.

As with many misconceptions, the "Source and Sink" idea has an element of truth to it. A battery does store energy, and the energy is used up in the lightbulb (or, more correctly, converted to heat and light in the lightbulb). In that sense, the battery is the source and the lightbulb is the sink, but for **energy,** not electric current.

Many teachers seek metaphors to help explain electricity. Among the most frequent are comparison to the flow of water in a plumbing system, and comparison to crowds moving through passageways. Both have their strengths and drawbacks. For more on this, a document can be downloaded from www.psych.northwestern.edu /psych/people/faculty /gentner/newpdfpapers /gentner-gentner-83.pdf.

In fresh water, ions are scarce, making it a rather poor conductor. Adding salt to the water makes it a much better conductor. Even fresh water, though, is a good enough conductor to be an electrical hazard, and the danger of mixing water and electric appliances should not be diminished.

Hopping Electrons
Misconception: "Electrons jump from atom to atom."

In a conventional conductor such as copper wire or other metal, some electrons help bind the atoms in a crystal structure. These electrons are fixed in place and do not flow. Other electrons are not fixed in place. They're not associated with any particular atom within the conductor, but move freely wherever the electric forces push or pull them. In other words, the electrons in a conductor are part of the material, but don't belong to a specific atom within the material.

A more accurate way to describe the current in a conductor is to say that the electrons move around *among* the atoms, often bumping into the atoms as they move (the main cause of electrical resistance).

Atoms and Electrons
Misconception: "Electrons are small parts of atoms."

Well, that's true except for the electrons that are *not* parts of atoms.

As mentioned above, the electrons in a conductor do not belong to a specific atom within the conductor. Electrons even exist on their own with no atoms around at all. These are called free electrons. In outer space there are free particles of many kinds, including electrons. You don't have to go to space to find free electrons. Vacuum tubes, TV picture tubes, and computer monitors work with currents of free electrons.

Electrical Inventors from Session 1

Thomas Edison (1847–1931)

"The Wizard of Menlo Park" (New Jersey) patented 1,093 inventions, including the storage battery, phonograph, microphone, telephone receiver, mimeograph, and motion-picture camera. Edison also created the world's first industrial research laboratory, and helped form the Edison Electric Light Company, the predecessor of the General Electric Company. He was a genius at assembling scientific talent and coaxing creativity from the teams he worked with.

Lewis Howard Latimer (1848–1928)

Lewis Latimer was the only African American member of Edison's team, and an excellent draftsman. He developed and patented the process for making carbon filaments. He also invented the threaded lightbulb socket, and patented a "water closet" (toilet compartment) for railway cars.

Grace Murray Hopper (1906–1992)

Grace Hopper was one of the first computer programmers in the country. Dr. Hooper worked hard to make computers more programmer- and application-friendly. She was a great leader in the field of software-development concepts, and an admired teacher.

Nikola Tesla (1856–1943)

Many of the new technologies in the world today are based on Nikola Tesla's ideas. This inventor, electrical engineer, and scientist invented the radio, fluorescent lights, the Tesla Coil transformer, rotating magnetic field principle, wireless transmission of electrical energy, alternating-current power transmission, remote-control boat, and more than 700 other patented devices or processes. His powerful contributions have been recognized with an International Unit of Magnetic Flux Density called the "Tesla."

Hertha Marks Ayrton (1854–1923)

Physicist Hertha Ayrton investigated the ripple movements in sand and water. She also worked on the behavior of the electric arc, and wrote what would become a standard textbook on the subject. She invented an instrument that divided a line into any number of equal parts and a fan to clear poisonous gases from trenches in World War I.

Granville Woods (1856–1910)

Granville Woods was a great electrician and an inventive genius. He invented 15 appliances for electric railways. He patented an improved steam boiler furnace and a telephone transmitter, and founded the Woods Electrical Company, which made and sold telephone, telegraph, and electrical instruments. Among his more than 35 mechanical and electrical inventions are a system for letting a train engineer know of trains in front and behind, the "third rail" used in subways, and the trolley-car system.

Otis Boykin (1920–1982)

Boykin invented more than 25 electronic devices, including a type of resistor used in computers, radios, and television sets. He also invented a control unit for an artificial heart stimulator (pacemaker), the electrical device used in all guided missles and IBM computers, a burglar-proof cash register, and a chemical air filter.

Mark Dean (1957–present)

Of the computer that all PCs are based on, Mark Dean holds three of the original nine patents. He and a coworker, Dennis Moeller, developed the ISA systems bus that allows multiple devices, such as a printer and modem, to be connected to personal computers. At the time of this printing, Dean is working on the "electronic tablet," a magazine-sized device that, among other things, may eventually replace newspapers, magazines, and books.

Alexander Graham Bell (1847–1922)

The 30 inventions of this famed Scottish scientist include the telephone and the photophone-transmission of sound on a beam of light, the precursor of fiber optics. He also invented methods to teach speech to the deaf.

Michael Faraday (1791–1867)

Like Benjamin Franklin, this English physicist was one of the most influential scientists in the field of electricity. Discovering that electricity could be made by moving a magnet inside a wire coil, Faraday was able to build the first electric motor and, later, the first generator and transformer. He introduced several words that we still use today to discuss electricity: *ion, electrode, cathode,* and *anode.* A unit of electricity was named after him; the "farad" measures an amount of electrical charge. Faraday is also remembered for his contributions in chemistry, especially his discovery of benzene.

Samuel Morse (1791–1872)

A portrait painter by profession, Morse became excited by the possibility of sending long-distance messages using electricity. In 1837 he patented the electric telegraph, the first electrical invention to have a major impact on society.

International Morse Code

A	di, dah • —	I	di, dit • •	R	di, da, dit • — •	0	da, da, da, da, dah — — — — —
B	da, di, di, dit — • • •	J	di, da, da, dah • — — —	S	di, di, dit • • •	1	di, da, da, da, dah • — — — —
C	da, di, da, dit — • — •	K	da, di, dah — • —	T	dah —	2	di, di, da, da, dah • • — — —
D	da, di, dit — • •	L	di, da, di, dit • — • •	U	di, di, dah • • —	3	di, di, di, da, dah • • • — —
E	dit •	M	da, dah — —	V	di, di, di, dah • • • —	4	di, di, di, di, dah • • • • —
F	di, di, da, dit • • — •	N	da, dit — •	W	di, da, dah • — —	5	di, di, di, di, dit • • • • •
G	da, da, dit — — •	O	da, da, dah — — —	X	da, di, di, dah — • • —	6	da, di, di, di, dit — • • • •
H	di, di, di, dit • • • •	P	di, da, da, dit • — — •	Y	da, di, da, dah — • — —	7	da, da, di, di, dit — — • • •
		Q	da, da, di, dah — — • —	Z	da, da, di, dit — — • •	8	da, da, da, di, dit — — — • •
						9	da, da, da, da, dit — — — — •

About the Components Used in the Unit

Wire

Wire sizes

The "gauge" of a wire describes its thickness. The higher the gauge, the thinner the wire. Single-strand telephone wire is usually 26 gauge, and is fairly thin and flexible. The wire used for standard home wiring is 14 gauge, and is very strong and stiff. Most appliance power cords use multi-strand wire, which combines many strands of narrow-gauge wire to create the equivalent of one wider-gauge wire. Multi-strand wire is flexible and easy to cut. This is the wire used in the holiday light strings students are given in the unit.

Bell wire

Bell wire is single-strand wire with plastic or rubber insulation. It's suitable for many low-voltage electric projects.

Magnet Wire

Magnet wire is single-strand copper wire. It's insulated with a tough thin lacquer coating. The lacquer may be colored (usually red, but sometimes blue or green), or the lacquer may be clear and colorless, making the wire look as if it has no insulation at all.

Electromagnets have wire that's wrapped in tight coils, usually around an iron core. The magnets are more compact and efficient when the wire can be wrapped closely and tightly. That's why magnet wire has thin lacquer instead of plastic or rubber insulation. (Think of the wire wrapping in an electric motor.)

Using magnet wire in the activities in this guide will work, but stripping it of its insulation adds complications. If you happen to have magnet wire available to you, consider making it available to students to increase the variety of materials they can use in their projects.

See "Stripping the wire" in "Acquiring and Preparing Materials" on page 193.

Holiday Light Strings

Modern strings of miniature holiday lights are made to operate on regular 120-volt "house current." Each individual bulb is designed to work at two and a half volts. In the light-string circuit there are generally 50 bulbs wired in series. These "share" the voltage, which is distributed evenly to each bulb. If the string has more than 50 bulbs, strings of 50 bulbs in series are wired together in parallel. That's why some of the sockets in the string have three wires coming from them instead of two. The connections between parallel circuits are in those sockets.

In the old days, when one bulb in the string blew out, every light in the string went out, and it was a chore to find which bulb in the string had died. Nowadays, the bulbs are made so that when the filament in a bulb burns out, a switch right inside the bulb closes, to keep the connection in the circuit. As a result, the burned-out bulb stays dark but the rest shine on. In fact, the rest shine a little brighter, because there are fewer bulbs sharing the voltage in the circuit, and therefore there's less resistance. Because the other bulbs shine more brightly, they burn out more quickly if the bad bulb isn't replaced.

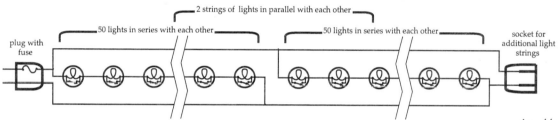

An abbreviated diagram of a typical string of 100 lightbulbs.

When several bulbs are burned out, the current through the string goes up and bulb burn-out becomes epidemic. You'll usually see a warning on the package to the effect that bulbs should be replaced promptly. If all the bulbs burn out, the light string becomes a short circuit, which would be dangerous. The manufacturers put a fuse into the plug to prevent this from happening.

Batteries

Sizes and shapes
What most people call a "battery" is, in fact, an electric cell. A true battery is two or more cells connected so that they work together. To be absolutely accurate, individual AAA, AA, D, and C "batteries" really should be called "cells." When more than one are connected, they become a battery. (Nine-volt batteries and car batteries really are batteries, because they contain more than one cell.)

The symbol for a battery is actually different from that for a cell too; here are two symbols used for true batteries:

For simplicity's sake, where more than two cells are pictured together in a circuit in this guide, we simply use two symbols for a cell rather than the symbol for a battery.

Batteries are described by their size, shape, and voltage. The difference between AAA-, AA-, C-, and D-size batteries is simply size and how long they last. They're all in the same voltage range. The larger batteries usually last longer. While students can incorporate any of these batteries into their inventions, the batteries recommended for these activities are C-size, because they come in the right voltage range (1.2 V to 1.5 V) and fit so well into a film-canister battery holder. *Note:* Nine-volt batteries (those small rectangular ones with the snap tops) are true batteries, with several electric cells built into them. They'll burn out the lightbulbs used in these activities very quickly unless you have three or more bulbs in series.

Regular, alkaline, and rechargeable
All of the batteries listed above come in regular, alkaline, and rechargeable form.

Regular batteries
These are usually the least expensive, but they have the shortest lifespan. When regular batteries are getting low, the voltage drops—making lightbulbs in a circuit shine more dimly.

Alkaline batteries
These last longer, but are more expensive. They tend to keep a steadier voltage than regular batteries, with the result that lightbulbs in a circuit don't go dim until the batteries have almost run out.

Rechargeable batteries
Rechargeables are the most expensive batteries to buy, but they can be used many times. They have slightly lower voltages than regular or alkaline batteries, so that lights in a circuit shine a little less brightly. Read the instructions on the battery recharger carefully. Batteries that are charged improperly sometimes run out quickly between rechargings.

Note: **Short circuits made with rechargeable batteries get very hot—much hotter than short circuits made with regular or alkaline batteries. For open-ended explorations in which students are likely to make inadvertent short circuits, we recommend that you do not use rechargeable batteries.**

How do they work?

Inside every battery is a combination of chemicals. These react with one another to create a flow of ions (charged atoms) inside the battery. This chemical reaction can only happen if there's an electric current outside the battery. When the battery is in a complete circuit, there's a flow of electrons outside the battery and a flow of ions inside the battery, making a complete closed loop of flowing electric charge.

Many different combinations of chemicals are used in batteries. Most of them use different kinds of metal separated by an acid (or a base, for an alkaline battery).

The principles of a battery were discovered accidentally by Luigi Galvani (1737–1798), an Italian physician who noticed that the legs of dead frogs would twitch when he touched them with his dissecting scalpel. The metal in the scalpel and the moisture in the frog legs made a battery. Many people with metal fillings have experienced an accidental "battery in the mouth." This happens when a piece of aluminum foil touches a dental filling that is damp with saliva. The resulting electric current is very unpleasant!

If electric current is forced backwards through a battery, several things may happen, depending on the type of battery. In a rechargeable battery this is a deliberate operation; the battery's chemical reaction goes in reverse, so that the battery is recharged and can be used again. In other batteries, a different kind of reaction, electrolysis, can occur. This reaction splits the water molecules in the battery and releases hydrogen gas. The buildup of gas can damage the battery or even cause it to explode. We recommend that you have students experiment with no more than two batteries in a circuit, so that, should a battery accidentally be inserted backwards, no more than one other battery can force current through it and cause a reaction.

Homemade Electric Devices

See the handouts on pages 198–219 for reproducible instructions on making the homemade devices used primarily in Sessions 2 and 3. ■

Lightbulbs

Finding and choosing light strings

All the projects in this unit use miniature lightbulbs from decorative "holiday" light strings. Hardware and variety stores often have these strings of lights for sale during December, and you may even find them *on sale* in early January. Many stores have them on display year 'round, but sometimes in the off-season you need to ask the store personnel to check in "the back room."

Although one string of 100 bulbs may be just enough for a class of thirty two students, there are a few good reasons to buy two strings. There's less chance of running short due to lost or burned out bulbs; you can be more generous with the number of bulbs the students may use in their projects; and you can cut some of the bulbs with longer wire leads, which makes them easier for students to work with.

It's best to use strings with all white, not colored, lightbulbs. The lights will shine quite dimly in some of these activities, and colored bulbs can make it difficult for the students to tell whether they're shining or not. Since students are allowed to use two batteries in their final projects, brightness is usually not a problem and you may want to have colored lights available. Among colored lights, yellow bulbs work best and blue bulbs work least well.

Light strings come in regular straight strings and in "icicle style." You may use either kind, but the straight strings are preferable because they usually have more extra wire in them, which you can use in these activities.

The farther apart the lights are spaced on the string, the better. Only occasionally will a manufacturer state the spacing on the package, but about three and one-half inches (eight or nine centimeters) is common. If you can find strings of lights spaced more widely, choose those.

Preparing the lights

1. Separate the strands.

There are usually two or three strands of wire twisted together in a string of lights. Only one of them has lightbulbs. Unravel the wires from each other and save the wires without lights for later use.

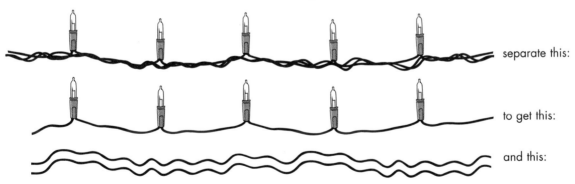

separate this:

to get this:

and this:

2. Cut the bulbs apart.

The bulbs are usually spaced too closely to cut every bulb with long enough leads. Cut the bulbs apart so that **every other bulb has 2 ½-inch leads.** This will give you a collection of bulbs with leads long enough for most of the activities, and a collection of bulbs with shorter leads that can be used in other activities in the guide.

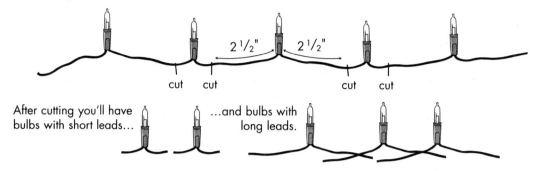

2 1/2" 2 1/2"

cut   cut   cut   cut

After cutting you'll have   ...and bulbs with
bulbs with short leads...   long leads.

3. Strip the leads.

Strip about ³/₄ inch of insulation off the end of each lead. Be careful not to tug too hard on the wire lead, or you may damage the bulb socket. (See pages 193–194 for more information about wire cutters and wire strippers.)

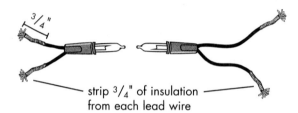

strip ³/₄" of insulation
from each lead wire

4. Harvest spare bulbs.

Almost all light strings have two or three sockets with **three** wire leads coming out of them. You should discard these, but save the bulbs as replacements for burned-out bulbs. A bulb can be removed by prying with your fingernails or pulling the small rim of plastic at the base of the bulb with one hand, while holding the socket with the other. **Don't pull on the wires or the glass bulb itself.**

save the bulb   discard the socket

Wire

Finding the wire

For the projects in this unit, there should be enough wire without bulbs in a string of lights. (See page 186 for information about different kinds of wire.) Wire may be reused from session to session. If you present these activities again or share materials with another teacher, you should have one set of wires for the more structured activities in Sessions 3, 4, 7, 8, and 9; you'll use these over and over again. You can then prepare some "consumable" wire to be used in the more creative activities in Sessions 5 and 10.

Stripping the wire

You, your students, and any other assistants you can enlist will need to strip many wires for these activities. Using the information that follows, simply cut wire to the desired length and strip about ³/₄ inch of the insulation from each end.

Wire strippers

Although a wire stripper is a fairly simple tool, it takes some practice to get comfortable using it. Some useful tips follow.

In Sessions 2 and 3, Station 6 is where students strip wire. Take advantage of the wire stripped by your students at this station for later sessions.

Adjusting your wire strippers

Most wire strippers work like short-blade scissors. The blades have a notch, so the stripper cuts the insulation (the plastic covering) but not the conductor (the wire). By then pulling the wire out from the blade, the insulation comes off. It sounds easy—and it is, if the size of the notch is adjusted correctly.

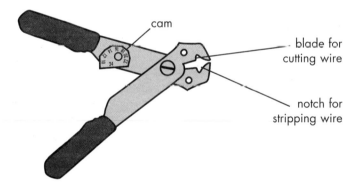

Many wire strippers have a screw or cam (a small unrounded wheel) that makes the notch larger or smaller. Try to adjust it so that the blade bites into the insulation without damaging the metal underneath.

Notch just right
Correct notching makes for easy wire stripping!

Notch too small
If the notch is too small, the blade can accidentally cut several strands from multi-strand wire. (For the activities in this guide, it doesn't matter if a few strands come off. Strands can come off even if the stripper is adjusted perfectly.) If you're using single-strand wire (such as bell wire), the wire may seem to strip fine, but you may have put a nick in the metal which will

weaken it. Then if someone bends the wire while making a connection, the end will break off.

Notch too big

This isn't as bad as a notch that's too small, but if the notch is too large the blade won't cut the insulation all the way. You usually can still strip the wire, but you'll have to work a little harder to break the small amount of uncut insulation as you pull it off. If the notch is *way* too big, the wire can slip out of the blade with the insulation still intact. You can sometimes twist the stripper as you pull, so that the blade digs into the insulation and successfully removes it.

Some wire strippers have several notches of different sizes instead of one adjustable notch. These are a little harder to use because you have to make sure the wire is in the right notch each time. This kind of stripper can be useful, though, if you're using wires of different sizes and don't want to keep readjusting your wire stripper.

Wire cutter versus wire stripper

Most wire strippers have a section of blade that cuts wire. If you have a good wire stripper, you don't need a separate wire cutter. Conversely, it's possible to strip wire using just a wire cutter, or even a pair of scissors, like this: Squeeze the handles *just hard enough* to cut through the plastic covering but **not** the wires inside, rotating the wire as you cut to create a groove all the way around. You should then be able to pull the insulation off from the end.

Paper Fasteners ("Brads")

Paper fasteners (often called "brads") come in a variety of sizes based on their shank size. In this unit, only brads with a $^3/_4$-inch shank are used. (A 1-inch shank is too big; a $^1/_2$-inch shank is too small.)

Throughout the unit the paper fasteners are primarily used to provide a secure metal-to-metal connection between the bare ends of wire and the battery *through* the slits in the film canisters. The $^3/_4$-inch shank on the paper fastener is just the right size to fit inside a film canister.

The paper fasteners are also used when constructing the homemade electric devices. They provide a "swivel-type" point of attachment or a metal-to-metal connection for the circuits in the devices.

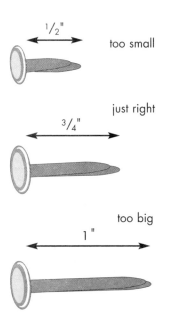

$^1/_2$" too small

just right $^3/_4$"

too big 1 "

Plastic Film Canisters

Finding the film canisters

Most facilities that process film save their used film canisters to be recycled. These facilities are usually willing to give them to teachers to be reused as educational raw materials. Depending on when the last recycling pick-up was, you may get several hundred containers at one stop!

Many large drug stores and variety stores process film right in the store. These stores are often the most accessible way to obtain film canisters. You can also check independent photo labs and stores, particularly those that advertise "one hour" or "same day" processing.

A Warning and a Suggestion

Sometimes in a bag of used film canisters you will find disposable cameras. Cameras with built in flash units have a high voltage source inside and can give a strong electric shock if they are taken apart. If you or another teacher are doing activities about optics (such as the *More Than Magnifiers* unit from GEMS), then used disposable cameras *that do not have built in flash units* can be used for students to take apart and explore.

Which film canisters will work?

In a bag of used film canisters you'll typically find Kodak and Fujifilm canisters and a few from less common brands. You'll find the traditional cylindrical canisters and the oval Advanced Photo System™ canisters. Only the cylindrical canisters with flat, snap-on lids will work for the projects using C-cell batteries described in this unit. This includes the Kodak canisters, but hardly any of the other brands.

Remember that one of the themes of this book is *inventiveness!* You or your students may invent good uses for all the canisters. There's probably a way to use Fujifilm canisters with AA batteries. The details are left to the teacher and/or the students to work out.

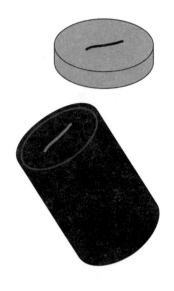

You may also choose to have others help you with the preparation of the raw material needed to construct the devices (such as taking apart the strings of holiday lights or stripping wires). If you do, be sure to copy pages 190–196 for these volunteers.

Preparing the film canisters

Each film canister must have a small slit in the center of the bottom and in the center of the lid. These slits can be made with a paring knife, utility knife, or pointed scissors. The slits should be made by an adult experienced in the safe use of the tools.

To make the slits, simply press the point of the tool carefully into the center of the bottom of each canister and the center of each lid. Make the slits about ¼ inch long.

Homemade Electric Devices

The task of making the six devices for Sessions 2 and 3 may seem daunting, but we've provided a variety of ideas to make it easier on you. The step-by-step, illustrated instructions that begin on page 198 have been written so the devices can be constructed by virtually anybody.

Options for making the devices

- **Have aides make them.** Have aides or parent volunteers make the devices for you.

- **Have families make them at home.** Send home to any willing families a container, or large resealable plastic bag, that includes the material and instructions to make one device. Ask them to construct them as a family, then have students bring them back to class in time to be used for Session 2. This option has the added benefit of inspiring families to do a small science project together!

- **Have students make them at school.** Have student volunteers make the devices during recess or after school.

- **Make them yourself.** Make all or some of the devices by yourself or with help, before starting the unit. If you choose this approach, we highly recommend teaming up with another teacher in your school, if possible, to share preparation and materials.

- **Make some, and add flashlights.** Make a few devices, and bring in dissected squeeze lights and flashlights students can take apart to substitute for the other devices. Students can try to figure out how the flashlights work, just as they would with the home-made devices (although we ask that they **not** take apart the homemade devices).

- **Don't make the optional devices.** We don't recommend this option, because it deprives the students of a critical learning experience and considerable engagement with the concepts. If absolutely necessary, however, skip the homemade devices the first time you present the unit, and use the devices your students design in Session 10 (Electric Inventions) as substitutes the next time you present the unit.

The directions that follow can be copied to give to helpers who make the devices for you. These are not the only, or even the best, ways to put the devices together. Feel free to modify or adapt them, but always use materials students will recognize and have access to when they're building their own projects.

Materials and Directions for Quiz Board

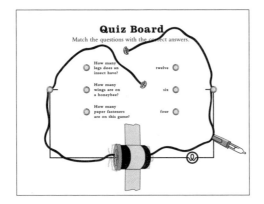

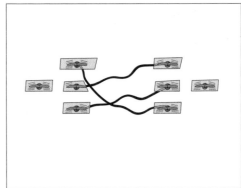

■ What You Need

- ❑ 1 copy of the **Quiz Board** template (page 220)
- ❑ 1 piece of stiff cardboard, 8 1/2 x 11 inches or slightly larger
- ❑ 1 film canister and lid with slits cut in top and bottom
- ❑ 12 paper fasteners (brads)
- ❑ 1 C-cell battery
- ❑ 6 pieces of 6-inch-long wire with stripped ends
- ❑ 1 holiday lightbulb with stripped leads
- ❑ glue or glue stick
- ❑ tape
- ❑ paring knife or scissors with sharp point
- ❑ *(optional)* clear contact paper

Unlike other masters for the teacher, the homemade-device templates do not have tabs identifying them as pages to photocopy. This is to minimize distraction for the students when the templates are in place.

■ Instructions

1. Glue the **Quiz Board** template to the cardboard rectangle.

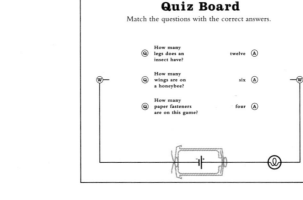

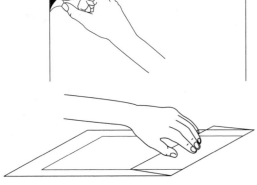

2. If you've decided to use it, apply clear contact paper to protect the template. Peel the backing off a sheet of contact paper that is larger than the cardboard. Place it sticky side up. Set the cardboard face down on the contact paper and fold the edges over.

3. Make a battery holder for a single C-cell battery as follows:

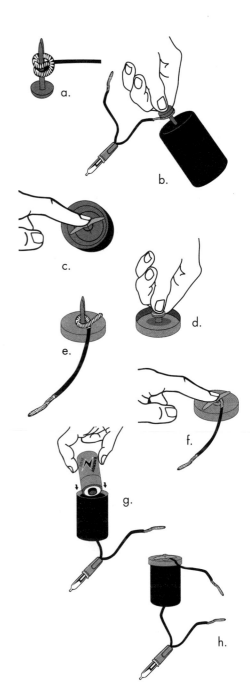

a. Wrap the stripped end of one of the wires from the lightbulb around the "legs" of a paper fastener just under the head.

b. Push the legs of that same paper fastener into the slit in the bottom of the film canister *from the outside to the inside*. The wire and the head of the fastener will remain outside—on the bottom of the canister.

c. Reach into the canister with one finger and separate the legs of the paper fastener. Push them apart, and press them firmly against the inside of the bottom of the canister.

d. Push the legs of another paper fastener through the slit in the lid of the film canister from the *inside to the outside.*

e. Wrap one stripped end of one of the wires without a bulb around the legs of the fastener right at the *top* of the lid.

f. Separate the legs of the fastener and push them firmly against the top of the lid.

g. Put the battery into the film canister with the bump-side (positive end) facing the bottom of the canister.

h. Fasten the lid on the canister. You will have to push hard around the edge of the lid so that it stays on firmly.

i. Test the circuit by touching the free ends of the wires together. The light should shine.

4. Tape the battery holder to the board over its picture.

5. Use the knife or scissors to poke small slits in the board at the holes marked with a "W". (An adult should do this.)

6. Wrap the stripped end of one of the wires from the battery holder around the legs of a paper fastener, and wrap the stripped end of a new wire around the same fastener. Insert the paper fastener through the nearest hole marked "W" and spread the legs of the paper fastener apart on the back of the board.

7. Do the same for the wire at the other end of the battery holder.

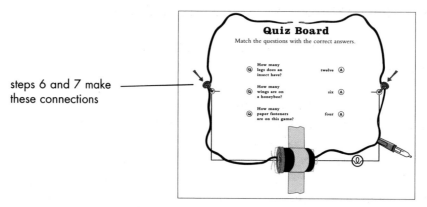

steps 6 and 7 make these connections

8. Wrap the stripped loose end of one of the wires that is connected to the board at a "W" around the legs of a paper fastener.

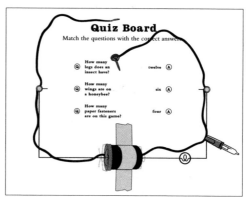

9. Separate the legs of the paper fastener and bend them up and over the head of the fastener. **This does not connect the wire to anything.** The purpose is simply to have the loose wire end in a small "lump" of metal.

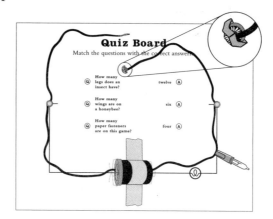

10. Do the same for the loose end of the other wire that's connected at the other "W".

11. Use the knife or scissors to poke small slits in the board at all the places marked with a "Q" or an "A". (An adult should do this.)

12. Poke a paper fastener through each of the "Q" and "A" slits from the front to the back.

13. On the back of the board, use a wire with stripped ends to connect the paper fasteners by the questions with the paper fasteners by the correct answers. [Insects have six legs, honeybees have four wings, and there are twelve paper fasteners in this device if you made it according to these instructions.] Connect the paper fasteners by wrapping the stripped end of the wires around them and pushing apart the legs of the paper fasteners.

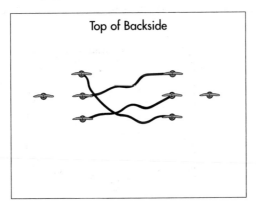

Top of Backside

14. Tape down all the legs of the paper fasteners on the back of the board. Make sure that the legs of one fastener do not touch the legs of the others.

15. Play the Quiz Board game to test the device.

Materials and Directions for Thread the Needle

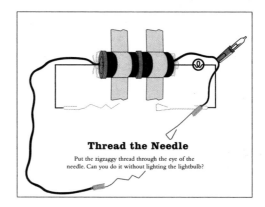

Thread the Needle

Put the zigzaggy thread through the eye of the needle. Can you do it without lighting the lightbulb?

■ What You Need

- ❏ 1 copy of the **Thread the Needle** template (page 221)
- ❏ 1 piece of stiff cardboard, 8 ½ x 11 inches or slightly larger
- ❏ 2 film canisters and lids with slits cut in tops and bottoms
- ❏ 3 paper fasteners (brads)
- ❏ 2 C-cell batteries
- ❏ 1 piece of 10-inch-long wire with stripped ends
- ❏ 1 holiday lightbulb with stripped leads
- ❏ 2 large paper clips
- ❏ 2 small pieces of aluminum foil, each 1 or 2 inches square
- ❏ glue or glue stick
- ❏ tape
- ❏ *(optional)* 1 pair of pliers
- ❏ *(optional)* clear contact paper

■ Instructions

1. Glue the **Thread the Needle** template to the cardboard rectangle.

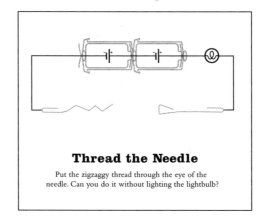

Thread the Needle

Put the zigzaggy thread through the eye of the needle. Can you do it without lighting the lightbulb?

2. If you've decided to use it, apply clear contact paper to protect the template. Peel the backing off a sheet of contact paper that is larger than the cardboard. Place it sticky side up. Set the cardboard face down on the contact paper and fold the edges over.

3. Make a double battery holder for two C-cell batteries as follows:

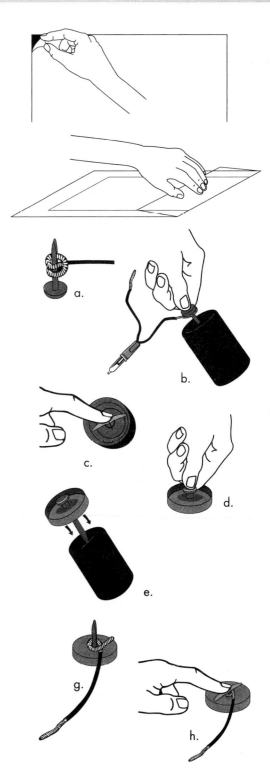

a. Wrap the stripped end of one of the wires from the lightbulb around the "legs" of a paper fastener just under the head.

b. Push the legs of that same paper fastener into the slit in the bottom of the film canister *from the outside to the inside.* The wire and the head of the fastener will remain outside—on the bottom of the canister.

c. Reach into the canister with one finger and separate the legs of the paper fastener. Push them apart, and press them firmly against the inside of the bottom of the canister.

d. Push the legs of another paper fastener through the slit in the lid of the film canister from the *inside to the outside.*

e. Attach this lid to the second film canister: Insert the paper fastener that pokes out of the top of the lid into the slit in the bottom of the second canister. Then reach into the canister, push the legs apart, and press them firmly against the inside of the bottom of the canister.

f. Insert another paper fastener into the lid of the second film canister, again from the inside to the outside.

g. Wrap one stripped end of the 10-inch wire around the legs of the paper fastener right at the top of the second lid.

h. Separate the legs of the fastener and push them firmly against the top of the lid.

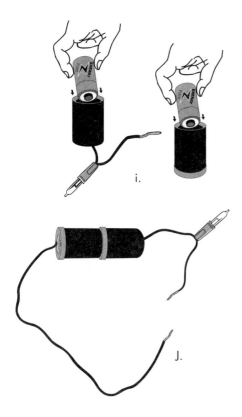

i.

j.

The "Needle"   The "Thread"

Foil and tape help make
a good connection
between the wire and
the paper clips.

i. Put a battery into each film canister with the bump-side (positive end) facing the bottom of each canister.

j. Fasten the lids on the canisters. Since one of the lids is attached to the bottom of a canister, you'll end up with two canisters joined together. You'll have to push hard around the edge of the lids so they'll stay firmly on the canisters.

k. Test the circuit by touching the free ends of the wires together. The light should shine.

4. Tape the battery holder to the board over its picture.

5. Bend a large paper clip so that it has a small, U-shaped loop at one end and a nearly circular loop, about $1/2$ inch across, at the other end. (The larger circular loop is the "eye of the needle.") Use pliers if necessary.

6. Wrap the stripped end of the remaining wire from the lightbulb around the U-shaped loop in the bent paper clip. Wrap the joint between the wire and the paper clip with a small piece of aluminum foil and use tape to hold it securely in place.

7. Bend the other large paper clip so that it has a small, U-shaped loop at one end. The other end should be mostly straight, with three or four bends to make a zigzag. (This is the "thread" that will go through the "eye of the needle.")

8. Wrap the remaining stripped end of the 10-inch wire around the U-shaped loop in the bent paper clip. Wrap the joint between the wire and the paper clip with a small piece of aluminum foil and use tape to hold it securely in place.

9. Test the device by touching the two paper clips together. The lightbulb should light up. Make sure that the clips are not touching when you store or carry the device.

Materials and Directions for Ice Fishing

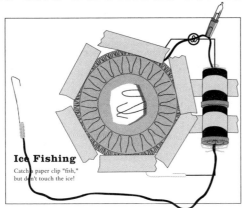

■ What You Need

- ❏ 1 copy of the **Ice Fishing** template (page 222)
- ❏ 1 piece of stiff cardboard, 8 ½ x 11 inches or slightly larger
- ❏ 2 film canisters and lids with slits cut in tops and bottoms
- ❏ 3 paper fasteners (brads)
- ❏ 2 C-cell batteries
- ❏ 1 piece of 10-inch-long wire with stripped ends
- ❏ 1 holiday lightbulb with stripped leads
- ❏ 1 miniature aluminum pastry dish, 4–5 inches across
- ❏ 1 large paper clip
- ❏ 6 small paper clips
- ❏ 2 small pieces of aluminum foil, each 1 or 2 inches square
- ❏ glue or glue stick
- ❏ tape
- ❏ *(optional)* 1 pair of pliers
- ❏ *(optional)* clear contact paper

■ Instructions

1. Glue the **Ice Fishing** template to the cardboard rectangle.

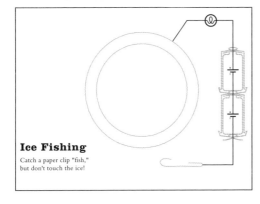

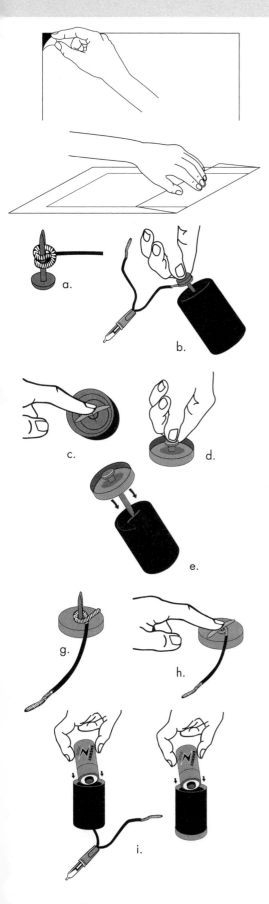

2. If you've decided to use it, apply clear contact paper to protect the template. Peel the backing off a sheet of contact paper that is larger than the cardboard. Place it sticky side up. Set the cardboard face down on the contact paper and fold the edges over.

3. Make a double battery holder for two C-cell batteries as follows:

   a. Wrap the stripped end of one of the wires from the lightbulb around the "legs" of a paper fastener just under the head.

   b. Push the legs of that same paper fastener into the slit in the bottom of the film canister *from the outside to the inside.* The wire and the head of the fastener will remain outside—on the bottom of the canister.

   c. Reach into the canister with one finger and separate the legs of the paper fastener. Push them apart, and press them firmly against the inside of the bottom of the canister.

   d. Push the legs of another paper fastener through the slit in the lid of the film canister from the *inside to the outside.*

   e. Attach this lid to the second film canister: Insert the paper fastener that pokes out of the top of the lid into the slit in the bottom of the second canister. Then reach into the canister, push the legs apart, and press them firmly against the inside of the bottom of the canister.

   f. Insert another paper fastener into the lid of the second film canister, again from the inside to the outside.

   g. Wrap one stripped end of the 10-inch wire around the legs of the paper fastener right at the top of the second lid.

   h. Separate the legs of the fastener and push them firmly against the top of the lid.

   i. Put a battery into each film canister with the bump-side (positive end) facing the bottom of each canister.

j. Fasten the lids on the canisters. Since one of the lids is attached to the bottom of a canister, you'll end up with two canisters joined together. You'll have to push hard around the edge of the lids so they'll stay firmly on the canisters.

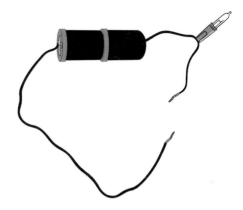

k. Test the circuit by touching the free ends of the wires together. The light should shine.

4. Tape the battery holder to the board over its picture.

5. Cut a hole about 2 or 3 inches across in the center of the bottom of the aluminum pan.

6. Wrap the stripped end of the remaining wire from the lightbulb around one end of a small paper clip.

7. Slide that small paper clip onto the edge of the aluminum pan. You may have to bend it a bit to fit the shape of the pan.

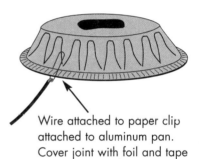

Wire attached to paper clip attached to aluminum pan. Cover joint with foil and tape to make a good connection.

8. To make sure that the electrical connection is solid, wrap a small piece of aluminum foil around the place where the wire, the paper clip, and the pan come together. Then use tape to hold it all in place.

9. Tape the aluminum pan upside down in the circle on the template. Tape all the edges down carefully, so nothing can slip through.

10. Bend the large paper clip so that it has a small, U-shaped loop at one end. Bend the other end into a broad U-shape. (This will form the fishing hook.) Use pliers if necessary.

Paper clip   Bent paper clip
fish hook   "fish"

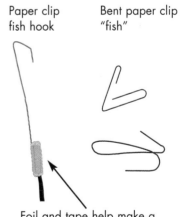

11. Wrap the remaining stripped end of the 10-inch wire around the small loop at the end of the bent paper clip. Wrap the joint between the wire and the paper clip with a small piece of aluminum foil and use tape to hold it securely in place.

12. Bend the wires of five small paper clips so that they do **not** lie flat when set down. Put them in the "hole in the ice" so that they rest on the board inside the aluminum pan.

Foil and tape help make a good connection between the wire and the paper clip.

13. Test the device by trying to fish the paper clips out of the pan with the hook. The lightbulb should light up only if you touch the edge of the "ice" (the pan) with the hook.

Materials and Directions for Count Up

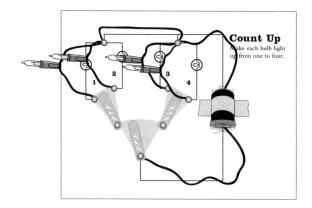

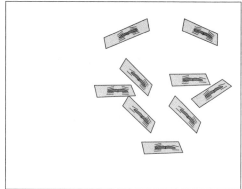

■ What You Need

- ❏ 1 copy of the **Count Up** template (page 223)
- ❏ 1 piece of stiff cardboard, 8 ¹/₂ x 11 inches or slightly larger
- ❏ 1 film canister and lid with slits cut in top and bottom
- ❏ 11 paper fasteners (brads)
- ❏ 1 C-cell battery
- ❏ 3 pieces of 6-inch-long wire with stripped ends
- ❏ 4 holiday lightbulbs with stripped leads
- ❏ 3 large paper clips
- ❏ 3 small pieces of aluminum foil, each 1 or 2 inches square
- ❏ glue or glue stick
- ❏ tape
- ❏ paring knife or scissors with sharp point
- ❏ *(optional)* clear contact paper

■ Instructions

1. Glue the **Count Up** template to the cardboard rectangle.

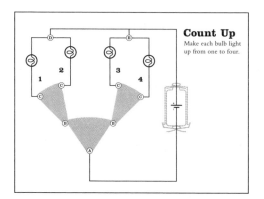

2. If you've decided to use it, apply clear contact paper to protect the template. Peel the backing off a sheet of contact paper that is larger than the cardboard. Place it sticky side up. Set the cardboard face down on the contact paper and fold the edges over.

3. Make a battery holder for a single C–cell battery as follows:

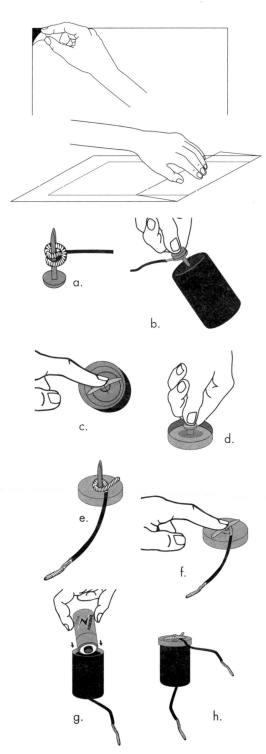

a. Wrap the stripped end of one of the 6-inch wires around the "legs" of a paper fastener just under the head.

b. Push the legs of that same paper fastener into the slit in the bottom of the film canister *from the outside to the inside.* The wire and the head of the fastener will remain outside—on the bottom of the canister.

c. Reach into the canister with one finger and separate the legs of the paper fastener. Push them apart, and press them firmly against the inside of the bottom of the canister.

d. Push the legs of another paper fastener through the slit in the lid of the film canister from the *inside to the outside.*

e. Wrap one stripped end of another 6-inch wire around the legs of the fastener right at the *top* of the lid.

f. Separate the legs of the fastener and push them firmly against the top of the lid.

g. Put the battery into the film canister with the bump-side (posi-tive end) facing the bottom of the canister.

h. Fasten the lid on the canister. You will have to push hard around the edge of the lid so that it stays on firmly.

4. Tape the battery holder to the board over its picture.

5. Use the knife or scissors to poke small slits in the board at all the places marked with an "A", "B", "C", "D", or "E". (An adult should do this.)

6. Wrap each of the three large paper clips with pieces of aluminum foil.

7. Poke the legs of a paper fastener though the foil at one end of one of the foil-wrapped paper clips. Push only the tips of the legs through for now.

8. Wrap the stripped end of the wire from the lid end of the battery holder around the paper fastener that's piercing the foil on the wrapped-up paper clip near the head of the fastener. Insert the paper fastener into hole "A", push it all the way through, and spread the legs on the back of the board.

9. Poke paper fasteners through the foil at one end of each of the other two foil-wrapped paper clips, insert the legs into the holes marked "B", and push them all the way through. Spread the legs on the back of the board. These paper clips do not have any wires attached to them.

10. Now you'll use paper fasteners to attach one of the wires from each of the four lightbulbs to the four holes marked "C". For each light-bulb, wrap one stripped end around a paper fastener, insert the fas-tener into hole "C", push it all the way through, and spread the legs on the back of the board.

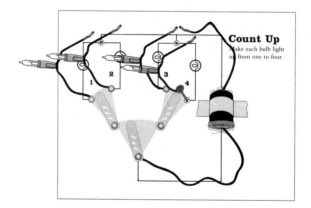

11. Wrap one stripped end of the remaining 6-inch wire around the legs of a paper fastener and then wrap the remaining wires from the lightbulbs marked "1" and "2" around the same paper fastener. Insert this paper fastener into the hole marked "D", push it all the way through, and spread the legs on the back of the board.

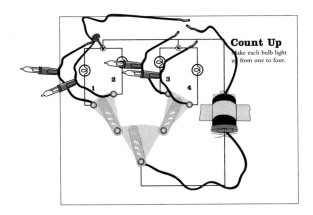

12. Now you'll wrap the stripped ends of four wires around the remaining paper fastener. Two of the wires come from the light-bulbs marked "3" and "4", one of the wires is the one that comes from the hole marked "D", and the fourth is the remaining wire from the bottom of the battery holder. Insert this paper fastener into the hole marked "E", push it all the way through, and spread the legs on the back of the board.

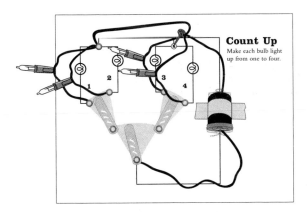

13. Tape down all the legs of the paper fasteners on the back of the board. Make sure that the legs of one paper fastener do not touch the legs of the others.

14. Test the device. The foil-wrapped paper clips should swivel around the paper fastener at one end so that the other end can touch the heads of other paper fasteners. When the foil-wrapped paper clips are pressed firmly against the heads of paper fasteners to make connections, one of the lightbulbs should shine.

Materials and Directions for What Matches?

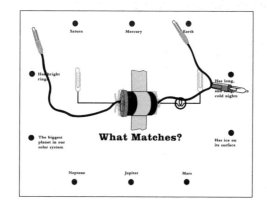

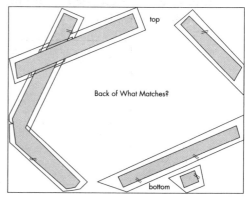

■ What You Need

- ❑ 1 copy of the **What Matches?** template (page 224)
- ❑ 1 copy of the **Back of What Matches?** template (page 225)
- ❑ 1 piece of stiff cardboard, 8 ½ x 11 inches or slightly larger
- ❑ 1 film canister and lid with slits cut in top and bottom
- ❑ 2 paper fasteners (brads)
- ❑ 1 C-cell battery
- ❑ 1 piece of 8-inch-long wire with stripped ends
- ❑ 1 holiday lightbulb with stripped leads
- ❑ 2 small paper clips
- ❑ several pieces of aluminum foil
- ❑ glue or glue stick
- ❑ tape
- ❑ a hole punch
- ❑ *(optional)* clear contact paper

■ Instructions

1. Glue the **What Matches?** template to the cardboard rectangle.

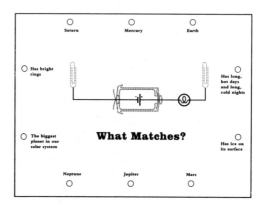

2. Turn the cardboard over and glue the **Back of What Matches?** template to the back, being sure to orient the top to what is the top on the front template.

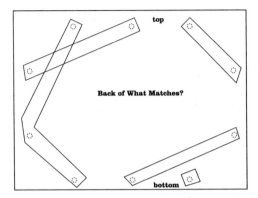

3. If you've decided to use it, apply clear contact paper to protect both of the templates. First, cut two pieces of contact paper—one slightly smaller than the cardboard and one slightly larger. Peel the backing from the slightly smaller piece and apply it to the back (over the back template).

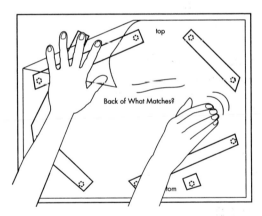

Peel the backing from the larger piece and place it on a table with the sticky side up. Set the cardboard face down on the contact paper (thus covering the front template) and fold the edges over to the back.

4. Use the hole punch to punch out each of the ten circles marked near the edge on the front template.

5. Now you'll make about six foil "wires." For each one, cut or tear a strip of foil about two inches wide and ten inches long. Fold each lengthwise in half twice to make a strong strip of foil about half an inch wide (and still ten inches long).

6. Following the outlines shown on the back template, put down the foil strips to connect the correct holes.

    a. For the one hole (Earth) that's connected to two other holes (Has ice on its surface, and Mars), use one full ten-inch strip of foil and bend it.

    b. For the other connections, tear off the lengths you need from the foil strips.

    c. Put tape between the place where two strips of foil cross so the metal of one strip doesn't touch the metal of the other.

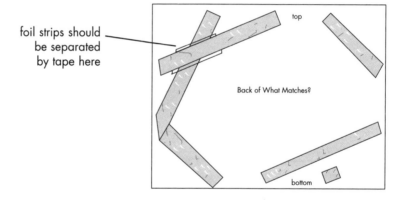

foil strips should be separated by tape here

top

Back of What Matches?

bottom

    d. One hole (for Neptune) is not connected to any other hole. Put foil behind it anyway, so that it doesn't look different from the others when viewed from the front.

    e. Cover **all** of the foil on the back of the board with tape. There should be no exposed metal.

    f. On the front of the board, the foil should show through each hole.

7. Make a battery holder for a single C-cell battery as follows:

a.

    a. Wrap the stripped end of one of the wires from the lightbulb around the "legs" of a paper fastener just under the head.

b. Push the legs of that same paper fastener into the slit in the bottom of the film canister *from the outside to the inside*. The wire and the head of the fastener will remain outside—on the bottom of the canister.

c. Reach into the canister with one finger and separate the legs of the paper fastener. Push them apart, and press them firmly against the inside of the bottom of the canister.

d. Push the legs of another paper fastener through the slit in the lid of the film canister from the *inside to the outside*.

e. Wrap one stripped end of the 8-inch wire around the legs of the fastener right at the *top* of the lid.

f. Separate the legs of the fastener and push them firmly against the top of the lid.

g. Put the battery into the film canister with the bump-side (positive end) facing the bottom of the canister.

h. Fasten the lid on the canister. You will have to push hard around the edge of the lid so that it stays on firmly.

i. Test the circuit by touching the free ends of the wires together. The light should shine.

8. Tape the battery holder to the board over its picture.

9. Wrap the remaining stripped end of the 8-inch wire around the end of a paper clip. Wrap the joint between the wire and the paper clip with a small piece of aluminum foil and use tape to hold it securely in place.

10. Do the same with the remaining wire from the lightbulb.

11. Test the device by poking the paper clips into the holes so that they touch the foil. If the holes are connected on the back by foil, the lightbulb will shine. [The "answers" are: Saturn has bright rings; Mercury has long, hot days and long, cold nights; Earth has ice on its surface; there is no "match" for Neptune; Jupiter is the biggest planet in our solar system; Mars also has ice on its surface.]

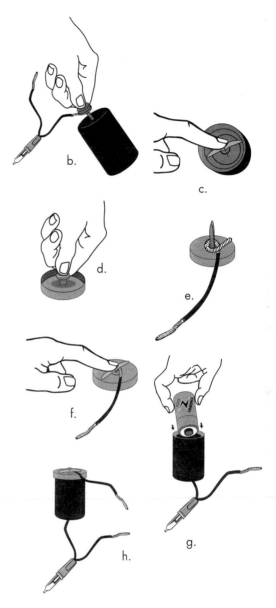

Materials and Directions for Upstairs/Downstairs

 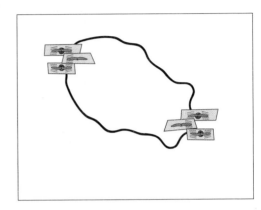

■ What You Need

- ❑ 1 copy of the **Upstairs/Downstairs** template (page 226)
- ❑ 1 piece of stiff cardboard, 8 ½ x 11 inches or slightly larger
- ❑ 1 film canister and lid with slits cut in top and bottom
- ❑ 8 paper fasteners (brads)
- ❑ 1 C-cell battery
- ❑ 3 pieces of 9-inch–long wire with stripped ends
- ❑ 1 holiday lightbulb with stripped leads
- ❑ 2 large paper clips
- ❑ 2 small pieces of aluminum foil, each 1 or 2 inches square
- ❑ glue or glue stick
- ❑ tape
- ❑ paring knife or scissors with sharp point
- ❑ *(optional)* clear contact paper

■ Instructions

1. Glue the **Upstairs/Downstairs** template to the cardboard
rectangle.

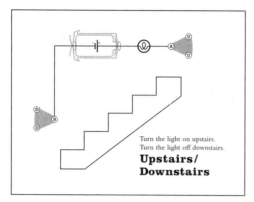

Turn the light on upstairs.
Turn the light off downstairs.

**Upstairs/
Downstairs**

2. If you've decided to use it, apply clear contact paper to protect the
template. Peel the backing off a sheet of contact paper that is larger
than the cardboard. Place it sticky side up. Set the cardboard face
down on the contact paper and fold the edges over.

3. Make a battery holder for a single C–cell battery as follows:

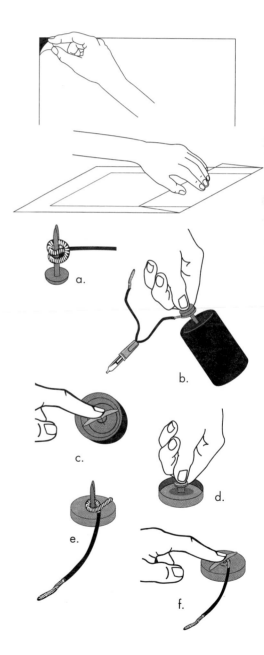

 a. Wrap the stripped end of one of the wires from the lightbulb
    around the "legs" of a paper fastener just under the head.

 b. Push the legs of that same paper fastener into the slit in the
    bottom of the film canister *from the outside to the inside*. The wire
    and the head of the fastener will remain outside—on the bottom
    of the canister.

 c. Reach into the canister with one finger and separate the legs of
    the paper fastener. Push them apart, and press them firmly against
    the inside of the bottom of the canister.

 d. Push the legs of another paper fastener through the slit in the lid
    of the film canister from the *inside to the outside*.

 e. Wrap one stripped end of one of the 9-inch wires around the
    legs of the fastener right at the *top* of the lid.

 f. Separate the legs of the fastener and push them firmly against the
    top of the lid.

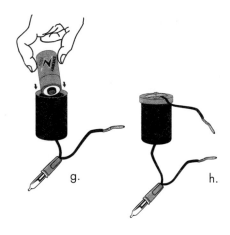

g.

h.

g. Put the battery into the film canister with the bump-side (positive end) facing the bottom of the canister.

h. Fasten the lid on the canister. You will have to push hard around the edge of the lid so that it stays on firmly.

i. Test the circuit by touching the free ends of the wires together. The light should shine.

4. Tape the battery holder to the board over its picture.

5. Use the knife or scissors to poke small slits in the board at all the places marked with an "A", "B", "U", or "D". (An adult should do this.)

6. Wrap each of the two large paper clips with pieces of aluminum foil.

7. Poke the legs of a paper fastener though the foil at one end of one of the foil-wrapped paper clips. Push only the tips of the legs through for now.

8. Wrap the stripped end of the remaining wire from the lightbulb around the paper fastener that's piercing the foil on the wrapped-up paper clip near the head of the fastener. Insert the paper fastener into hole "A", push it all the way through, and spread the legs on the back of the board.

9. Poke the legs of another paper fastener though the foil at one end of the other foil-wrapped paper clip. Push only the tips of the legs through for now.

10. Wrap the stripped end of the wire from the lid end of the battery holder around the paper fastener on the remaining foil-wrapped paper clip near the head of the fastener. Insert the paper fastener into hole "B", push it all the way through, and spread the legs on the back of the board.

11. Insert the remaining paper fasteners into the four holes marked "U" and "D".

12. On the back of the board, use one of the 9-inch wires to connect one of the "upstairs" paper fasteners to one of the "downstairs" fasteners. Wrap the stripped ends around the legs of the paper fasteners, spread the legs, and push them down flat. Use the remaining wire to connect the other pair of "upstairs/downstairs" paper fasteners, then spread and flatten the legs.

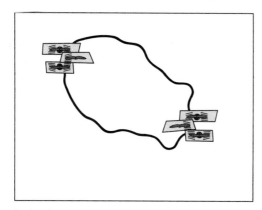

back of Upstairs/Downstairs

13. Tape down all the legs of the paper fasteners on the back of the board. Make sure that the legs of one paper fastener do not touch the legs of any others.

14. Test the device. The foil-wrapped paper clips should swivel around the paper fastener at one end so that the other end can touch the heads of the other paper fasteners. When the foil-wrapped paper clips are pressed firmly against the heads of the paper fasteners that are connected on the back, the lightbulb should shine.

Quiz Board

Match the questions with the correct answers.

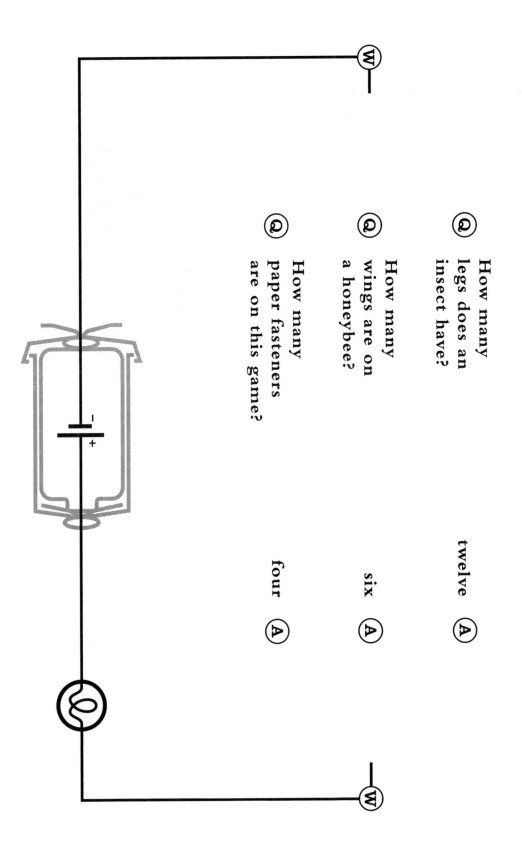

Q How many
legs does an
insect have?

twelve **A**

Q How many
wings are on
a honeybee?

six **A**

Q How many
paper fasteners
are on this game?

four **A**

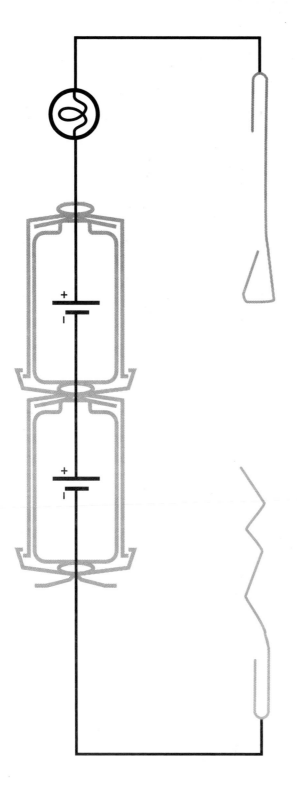

Thread the Needle

Put the zigzaggy thread through the eye of the needle. Can you do it without lighting the lightbulb?

Ice Fishing

Catch a paper clip "fish," but don't touch the ice!

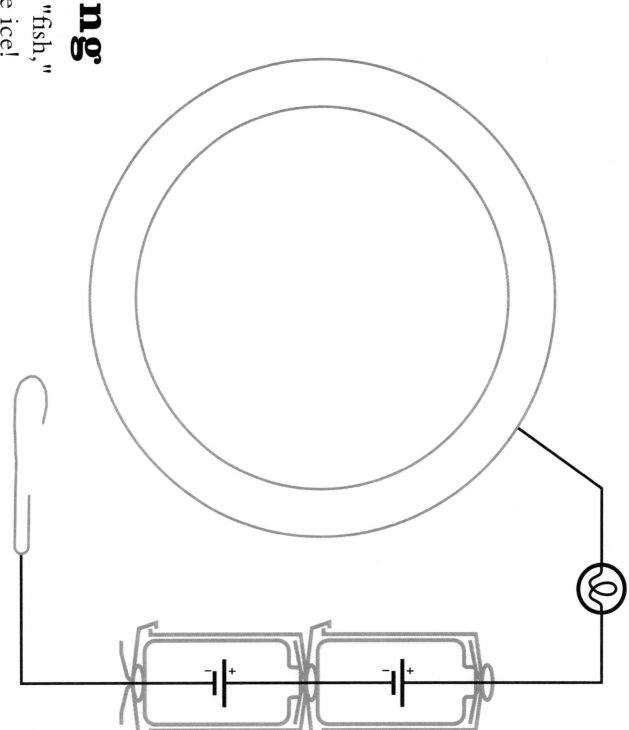

Count Up

Make each bulb light up from one to four.

What Matches?

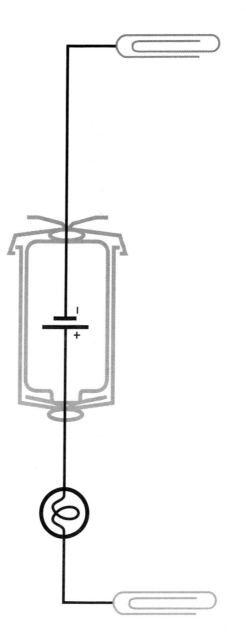

○ Has bright rings

○ Saturn

○ Mercury

○ Earth

○ The biggest planet in our solar system

○ Neptune

○ Jupiter

○ Mars

○ Has long, hot days and long, cold nights

○ Has ice on its surface

top

bottom

Back of What Matches?

Upstairs/
Downstairs

Turn the light on upstairs.
Turn the light off downstairs.

TEACHER'S OUTLINE

SESSION 1: INTRODUCING ELECTRICITY

■ Getting Ready
1. Prepare lightbulbs, wires, and film canisters. Read "Acquiring and Preparing Materials" on page 190.
2. Copy, color in, and post **No 120V** poster.
3. Set up overhead projector and screen.
4. Make and set out transparencies.
5. Copy **Electrical Inventor's Journal** for all students.
6. Set out one battery and one lightbulb with stripped leads for each pair of students. Have plug cut from light string ready to show during safety talk.

■ Edison's Journal
1. Project **Electrical Inventors** transparency and ask if students recognize any—they are inventors of electric devices as are students in this unit.
2. Project **Mystery Inventor's Journal** Transparency **1**. As you read aloud, challenge students to figure out the invention.
3. Repeat with remaining **Mystery Inventor's Journal** transparencies.
4. Ask why invention of the lightbulb was such a big deal and discuss.
5. Project transparency of **Lewis Howard Latimer**—an Edison team member.
6. Edison's team did lots of "messing around" before they succeeded—testing thousands of ideas and materials.
7. Students will also start by "messing around" with wires, lightbulbs, batteries. Later they'll make inventions.

■ Introducing the Electrical Inventor's Journals
1. Students will keep scientific journals. Let them know your expectations.
2. Distribute journals. Have them write what they know and wonder about electricity.
3. **Give safety information as in guide.**

■ Circuits Small and Simple
1. Organize students in pairs.
2. Their challenge is to light bulb, **using only battery and lightbulb.** Define *circuit.* When successful, students will draw device in journal.
3. Pass out materials and let pairs discover how to light bulb.
4. Circulate to make sure everyone has a chance to use materials.
5. As students finish, review what a circuit is, and check for basic understanding.
6. Collect materials; have students describe how they made bulbs light.

SESSION 2: ELECTRIC EXPLORATION STATIONS

■ Getting Ready
1. Prepare devices according to instructions.
2. Use diagrams in guide to plan station set up.
3. Prepare stations as described in the guide.
4. Copy **Electric Exploration Stations** journal pages for all students.

■ Introducing the Electric Exploration Stations
1. Student pairs will rotate through stations, but make individual journal entries. Tell them they must visit Stations 1 and 2 now or in next session.
2. They will need to read station signs and look at illustrations.
3. Discuss any difficult stations without doing activities or revealing results. Review concept of circuit, if needed.
4. Tell students to **disconnect circuits when they leave station.**

■ Stations Overview
1. For each station, summarize and explain as in guide.
2. For the repair shop, emphasize **students SHOULD NOT try to repair 120-volt electric devices!**

■ Exploring the Stations
1. Remind students of points to remember as they go through stations, as in guide.
2. Organize students into teams and pass out journal pages.
3. Assign teams to first stations and have them begin.
4. Five or ten minutes before class ends, have them write/draw in journals.
5. Leave stations for use in next session. Make sure all circuits are disconnected.

SESSION 3: SIMPLE CLOSED CIRCUITS WITH STATIONS

■ Getting Ready
1. Copy **Battery, Lightbulb,** and **Wire or Other Conductor** posters.
2. Make sure stations are in working order.
3. For each student, strip leads of 20-inch wire.
4. Set out materials as in guide.
5. If posting key concepts, write them out.

■ How the Electric Devices Work
1. Have students remember their previous work at stations. Ask what makes the electric devices work.
2. Ask them to name objects current does and does not flow through; list objects. If there's disagreement, one or two students could test.
3. Ask what scientists call such materials. Write "Conductors" at top of first column of list on board.

4. Ask what conductors have in common.
5. Further define conductor as in guide. Say that some metals, the ground, and water are good conductors. Add them to list.
6. Ask what materials that do not carry electric current are called. Write "Insulators" at top of second column.
7. Add "your body" to list of conductors. A strong electric current passing through a body can cause severe burns or death.
8. Remind them low voltage of batteries makes these classroom activities safe.
9. At Station 3, students will list conductors and insulators in their journal.

■ Creating a Class Circuit
1. Have students stand in circle. Have one hold a battery and one on opposite side hold a lightbulb.
2. Ask what's needed to light the bulb *without* bringing bulb and battery together.
3. Pass out wire to every student *except* those holding bulb and battery.
4. Have students hold or twist wires together as described in guide.
5. Emphasize activity requires teamwork; help them work together until bulb lights.
6. Say they've made a *large* circuit, but it's still called a "simple circuit" and they'll learn more about it. Keep students in circle for following discussion.

■ Introducing Electric Current and Circuits
1. Explain that shining bulb shows there's an *electric current* flowing in wire.
2. One way to understand is to **follow the path the current takes.** "Following current" is an important strategy.
3. With students' help, follow electric current through circle. Its path ends where it starts. This path of a current is called a *circuit.*
4. It's called a circuit even when a connection is undone and current doesn't flow—then it's an *open circuit.* A complete circuit with unbroken path for current is a *closed circuit.* Have students open and close the circuit and identify it as open or closed by whether or not the light is shining.

■ More Voltage, More Current
1. How could you increase the current to make light shine brighter?
2. Step into circuit between two students with wires and insert another battery. Discuss and demonstrate direction of battery's ends as described in guide.
3. Say that adding more batteries increases the power, or *voltage,* in the circuit. Collect equipment, dismiss students, and thank them for working together.

■ More Important Safety Messages
1. Tell students activity illustrates an important message—circuits with **more than two batteries** (or **different kinds of batteries**) should *never* have batteries installed pointing in opposite directions. Explain why as in guide.
2. Draw attention to the **No 120V** poster, and review what it means.
3. Ask what could happen to a person's body from 120 volts (or more).
4. Ask those who've been to Shocking Stories station to share safety tips. Mention the tips listed in guide.

■ Schematic Diagrams

1. Scientists use "picture language" to draw circuits—*schematic diagram.*
2. Hold up battery then draw schematic symbol. Do the same for bulb. Discuss each as in guide.
3. Ask student to finish connecting symbols to make simple closed circuit. Discuss that the lines connecting bulb and battery symbols represent any conductor.
4. Post posters for students' reference.
5. On board make simple drawing of one of the devices from the stations. Discuss the path current takes through the circuit in real device as in guide.

■ The Circuit Hunt: More Electric Exploration Stations

1. Students will go to stations they haven't been to yet.
2. Also, they'll be on "circuit hunt"—following current from battery, through all parts of circuit, and back to battery. They'll draw circuits on illustrations in journals.
3. Refer to simple drawing of device on board. Trace circuit through device as in guide.
4. Erase chalk mark through battery and ask, "Is this still a drawing of a closed circuit?" Draw line back in and say students need to draw complete loops to represent current in circuit.
5. Emphasize if students haven't completed Stations 1 and 2, they should do so first. Partner students as before, pass out journals, and have teams begin exploring stations.
6. Circulate and check that students are on task as listed in guide.
7. If you've chosen to post key concepts, display, discuss, and post them.

SESSION 4: DIAGRAMS AND SHORT CIRCUITS

■ Getting Ready

1. Copy **No Short Circuit!** poster.
2. Make **Light the Bulb** and **Building Circuits from Diagrams** transparencies. Copy **Experimenter's Kit List** and cut into strips.
3. Set out materials for "Experimenter's Kits" as in guide, along with one battery for each team of four.
4. Gather circuit device, lightbulb, and battery to debrief stations and demonstrate a short circuit. Have homemade devices available for debrief.
5. Copy **Building Circuits from Diagrams** sheets for students. Have them and journals ready to distribute.
6. If posting key concepts, write them out.

■ Debriefing Stations

1. Debrief Build a Circuit (#2) and Light the Bulb (#5) stations as in guide.
2. Ask if students fixed something at repair shop and for repair tips.
3. As you debrief remaining stations as in guide, hold up each device and model use of terms and concepts.

■ Discussing Short Circuits

1. Draw schematic from guide on board.
2. Ask students to follow current with you.

3. How is circuit different from others used?
4. Explain that when electricity can flow around a circuit *without* making a light shine—or anything else happen—it's a ***short circuit.***
5. Say the current in a short circuit can be very hot; even the heat from a battery could cause discomfort or minor burn.
6. What could happen if there's a short circuit with 120 volts or more? Have them share what they learned from Shocking Stories station.
7. Make sure they know that **short circuits can cause burns and fires, and cause electric devices to stop working.** They are also wasteful; battery drains.
8. They should try *not* to make short circuits; if it happens by accident, disconnect immediately. Show and post **No Short Circuits!** poster.

■ Examples of Short Circuits
1. Recount anecdotes about accidental short circuits as in guide.
2. Take a circuit device students made and reinsert battery.
3. Without doing it, show two ways to complete circuit without touching loose wires together. Draw schematic diagrams of both ways. Which shows a short circuit?
4. Say diagrams can reveal problems with a circuit before it's put together, or explain why a circuit doesn't work after it's put together.

■ Building Circuits from Diagrams
1. Show **Building Circuits from Diagrams** transparency. Go through each step (predict, build, write about, and explain) making sure students understand them.
2. Point out that students will not build circuit C and that they will design their own circuit in G.
3. Divide students into teams. Pass out data sheets, kit lists, batteries, and journals.
4. Have students assemble kits then make a battery holder as in guide. Have students begin building circuits. Circulate to check that students are engaged and recording in journals.
5. Ten or 15 minutes before class ends, have students remove batteries and reassemble kits. Collect batteries and kits; students keep journals for reference during class discussion.

■ Debriefing the Circuit-Building Activity
1. As in guide, ask series of questions about each circuit to review and stimulate thinking. If class will do Sessions 10 and 11, remind students they'll be designing their own devices and should record ideas now.
2. Say that in next session they'll invent switches.
3. If you've chosen to post key concepts, display, discuss, and post them.

SESSION 5: INVENT A SWITCH

■ Getting Ready
1. Copy **Switch** poster.
2. Gather circuit devices and batteries from Station 2 and stripped wires from Station 6. Also gather tape or stickers.
3. Designate a supply area and stock it with extra switch-building material.
4. Copy **Switch Invention Spec Sheet** for each student. Set out journals.

■ Introducing Switches
1. Hold up circuit device students made at Build a Circuit station. Insert battery and touch wires together to close circuit and light bulb.
2. On board, draw schematic diagram for simple closed circuit. Review parts (battery, lightbulb, wire) of diagram.
3. Say most circuits have something to make them closed circuits when current is to flow, and open circuits when it is to stop—most have *switches.*
4. Post **Switch** poster and add switch symbol to diagram.
5. Open and close switch in drawing, pointing out "on" means circuit is closed and current can flow; "off" means circuit is open and no current flows.
6. Hold up circuit device again, pointing out that when light is on, circuit is closed; when off, circuit is open.
7. Use switches in classroom to illustrate same concept—on means closed circuit; off means open circuit.
8. Have students list devices (as in guide) with switches, and discuss ways they work.

■ Inventing Switches
1. Look again at circuit device and say loose wires make a switch, but a crude one.
2. Tell class partners will invent their own switches. They must also make a "spec" sheet. Define spec sheet, show **Switch Invention Spec Sheet,** and tell what it must include.
3. Show supply area and materials students will work with. Pass out spec sheets.
4. Give pairs a few minutes to plan, then pass out batteries, circuit devices, and tape/stickers and have them begin.
5. When most have finished, call on a few to share problems and solutions.
6. Fifteen minutes before end of class, warn students they'll stop soon. Say they'll have a few minutes in next session to finish inventions and spec sheets. Ten minutes before end, have them stop and label switches and spare parts. Have students remove batteries from switch inventions; collect batteries.

SESSION 6: ELECTRIC SWITCH TRADE SHOW

■ Getting Ready
1. Set out materials from Session 5 so students can finish. Also set out journals.
2. If using, write "Rover's Recording Sheet" on paper and copy for each student.

■ Finishing Touches

1. Tell students about "trade shows." Say they'll have a trade show to share their switch inventions.
2. Before the show, each pair needs a finished switch and spec sheets.
3. Allow 15 minutes for students to finish switches and spec sheets.

■ The Trade Show

1. Divide each pair into Rover and Explainer. Define the role of each and say they'll switch halfway through the activity. If using them, distribute a "Rover's Recording Sheet" to each pair.
2. Let Rovers rove for 10 minutes, then have partners change and continue 10 minutes more.

SESSION 7: SERIES CIRCUITS

■ Getting Ready

1. Copy **Series Circuits** sheet for each student.
2. Set out journals, kits, and batteries.
3. Have extra wires, lightbulbs, paper fasteners, film canisters, and wire strippers available.
4. If posting key concepts, write them out.

■ Building Series Circuits

1. Tell students they'll make circuits with more than one bulb.
2. Pass out journals and **Series Circuits** sheet. Organize students into pairs. Have them look at first two circuits on page and predict which will have brighter lights.
3. Tell students they'll build **first two circuits** in teams of four; each student will write observations in her own journal.
4. After making first two circuits, teams may try others on sheet. Say they'll need to take first two circuits apart to have enough material for others.
5. If they build circuits that don't have diagrams, draw their own diagrams on back of sheet, noting how brightly lights shine.
6. Remind them to **avoid making short circuits.**
7. Organize students into teams. Pass out kits and batteries; say you have extra material. Have them begin.
8. After about 20 minutes, have students remove batteries and hand them in.
9. Have teams keep circuits intact, without batteries, for discussion.

■ Analyzing Series Circuits

1. Draw the circuit with one battery and two bulbs in series on the board.
2. With class, follow the current. Say it flows through "one bulb after another." Explain that this is what *series* means. Tell them the circuits they just built are *series circuits.*
3. Have volunteers describe how bulbs shine in circuit with one bulb or a circuit with two bulbs in series.
4. Draw schematic of circuit with one bulb next to circuit already on board.
5. Ask students how they might guess which circuit has more current flowing in it. Confirm that the brighter the light, the more current. A dimmer light means there's less current.

6. Tell them the conductor in a lightbulb is a long, thin wire called the *filament.* It is tightly coiled and longer than it looks. Say it's harder for current to flow through such a long, thin wire than to flow through shorter, thicker wire.

7. Again follow current around each circuit, but when you get to a bulb, model that it takes real effort for the current to push through.

8. Say anything that reduces flow of current has *resistance.* Each bulb adds more resistance. If circuits had buzzers, motors, or heaters, these would also add resistance.

9. Explain that while resistance of filament in bulb reduces flow of current, it is what makes the bulb shine. The current flowing through the filament heats the wire so much it glows brightly.

10. Did any teams build a circuit with one battery and **three bulbs in series?** What did they notice? Confirm that lights were very dim, or didn't shine.

11. Review what an open circuit is. Say an **open circuit** has so much resistance no current can flow. Establish that a short circuit has almost no resistance, so a battery will be drained quickly and heat will be generated.

12. Ask why lights in a short circuit don't shine extra bright. Confirm that current doesn't flow through bulbs in a short circuit.

13. Say with one battery, more resistance results in less current. Is there a way to increase current? Confirm that adding batteries will do it. Have teams that made series circuits with two batteries demonstrate them.

14. Have students disassemble circuits and put materials back into kits. Collect these, batteries, and journals. Tell students that in next session they'll put circuits with more than one bulb together in a different way.

15. If you've chosen to post key concepts, display, discuss, and post them.

SESSION 8: PARALLEL CIRCUITS

■ Getting Ready
1. Copy **Connection/No Connection** poster.
2. Copy **Parallel Circuits** sheet for each student.
3. Set out journals, kits, and batteries.
4. Have extra wires, lightbulbs, paper fasteners, film canisters, and wire strippers available.
5. Gather Quiz Board and What Matches? devices.
6. If posting key concepts, write them out.

■ Series Circuit and Parallel Circuit: Comparing Schematic Diagrams
1. Pass out journals and **Parallel Circuits** sheet.
2. Draw students attention to "two bulbs in parallel" circuit. Ask what it has that other circuits they've built didn't have.
3. Say where two or more conductors are connected, diagram shows a large spot. This is symbol of a connection that current can flow across.
4. Hold up Quiz Board and What Matches? as examples of circuits with **conductors that cross but do not connect.** Make sure students see the back, where conductors cross.

5. Post poster. Explain the two symbols as in guide.

6. Say whenever students draw a diagram showing conductors that come together, they should use the spot or the arch to make clear whether or not there's an electric connection.

■ Building the Series and Parallel Circuits

1. Put students in teams. Tell them they'll build the circuits in part A of sheet. Point out they'll have to take apart the first circuit and combine parts to have enough material for the second circuit, so they'll need to describe brightness right after building first circuit.

2. Be sure they know that after they build circuits in part A, they should **remove a bulb from each circuit** to see what happens.

3. Demonstrate how to remove a bulb. **Don't pull on the wires or the glass bulb itself.**

4. Tell teams that if they finish parts A and B, they may work ahead to do part C. Say if they don't get to it now, they'll have time after the class discussion.

5. Remind students to **avoid short circuits.**

6. Pass out batteries and kits; say you have extra material. Have teams begin.

7. Circulate and help teams follow the diagram.

8. When teams have completed parts A and B, collect batteries and prepare to start class discussion.

■ Comparing Parallel and Series Circuits

1. On board, draw diagrams for first two circuits students built.

2. Ask students to describe brightness of bulbs in each.

3. Ask why bulbs in series were dim. Say in a series circuit there's a single path for current to flow in.

4. Why were bulbs in other circuit brighter? Help them see that in second circuit, current goes two ways—and either way, it flows through only *one* bulb.

5. Formally introduce *parallel circuit* as a circuit with a place where the current splits or branches. Say in a parallel circuit there's more than one path for current to flow in.

6. A parallel *circuit* is different from parallel *lines*—in a circuit the "side-by-side" conductors *always* meet, and in more than one place. Point this out in the schematic diagram.

7. To review part B of the sheet, erase a bulb in each diagram on board to simulate its removal. Why does remaining bulb in the series circuit go out, while remaining bulb in parallel circuit stays on?

8. Are bulbs in classroom (or their homes) in series or parallel circuit?

■ More Explorations

1. Have a team that's built a complicated circuit demonstrate, diagram, and explain it to class.

2. There are many other circuits they can build with the material in kits.

3. Remind students to watch out for short circuits and to disconnect them immediately.

4. Redistribute batteries and have teams work on part C. Just before the end of the session, let some teams demonstrate and draw their parallel circuits.

5. Have teams disassemble their circuits and put materials back into kits. Collect these, batteries, and journals.

6. If you've chosen to post key concepts, display, discuss, and post them.

SESSION 9: SWITCHES IN SERIES AND IN PARALLEL

■ Getting Ready
1. Copy **Switches in Series and Parallel Circuits** sheet for each student.
2. Set out journals, kits, and batteries.
3. Have extra wires, lightbulbs, paper fasteners, film canisters, and wire strippers available.

■ Exploring Switches
1. Distribute journals and **Switches in Series and Parallel Circuits** sheet.
2. Point to **Switch** poster and remind students of the symbol.
3. Put students in teams. Assign a challenge from the sheet to each team. If they're successful at that challenge, they may choose another.
4. Say students don't have to make fancy switches; use wires they connect and disconnect.
5. Remind them to avoid short circuits and disconnect if it happens.
6. Pass out batteries and kits; say you have extra material. Let teams work for about 10 minutes.

■ Discussing Switches
1. Call on teams to describe and demonstrate how they tackled the challenges.
2. As necessary, describe possible solutions to challenges as in guide.
3. After the class discussion, have teams disassemble their circuits and put materials back into kits. Collect these, batteries, and journals.

SESSION 10: ELECTRIC INVENTIONS

■ Getting Ready
1. Copy **Electric Invention Assignment Sheet** and **Invention Spec Sheet** for each student.
2. Set out journals and homemade electric devices.
3. Gather stripped wire from Station 6.
4. Designate a supply area and stock it with extra building material.
5. Assemble batteries, film canisters, scissors, and tape or stickers.

■ Inventing Electric Devices
1. Pass out assignment sheets. Lead brainstorm of projects. Record their ideas on board.
2. If students have trouble coming up with ideas, share some as in guide. Remind them they can also get ideas from notes made in journals throughout the unit.
3. Pass out journals and spec sheets.
4. Emphasize that an important part of the project is their description of the *circuit* in their device. Remind students of the switch spec sheet, and if necessary review what a spec sheet is.
5. Go through the **Invention Spec Sheet** to make sure students know what they need to record. Point out they must include a schematic diagram and write about how current flows in the circuit.

■ Constructing Their Inventions

1. Organize students into pairs. Distribute material and show students the supply area.
2. Remind students to watch out for short circuits, and tell them they're limited to two batteries in their inventions.
3. Have pairs spend first few minutes making plans.
4. Have pairs begin working. Circulate and assist as needed.
5. Twenty minutes before end of class, warn students they'll stop soon. Say they'll have a few minutes in next session to finish inventions and spec sheets.
6. Fifteen minutes before end, have them stop and label inventions and spare parts. Have students remove batteries from inventions; collect batteries.

SESSION 11: ELECTRIC INVENTIONS FESTIVAL

■ Getting Ready

1. Set out materials from Session 10 so students can finish. Also set out journals.
2. If using, write "Rover's Recording Sheet" on paper and copy for each student.

■ Finishing the Inventions

1. Tell students that after they've finished their inventions, they'll describe and demonstrate them for the class.
2. Pass out journals and inventions, and give pairs five minutes to finish spec sheets and discuss what remains to be done on invention.
3. Pass out batteries, remind students about supply area, and give about ten more minutes to finish inventions.
4. Call on each pair to briefly describe their invention and demonstrate how it works.

■ Invention Festival

1. Divide each pair into Rover and Explainer. Remind students about role of each and that they'll switch halfway through. If using them, distribute a "Rover's Recording Sheet" to each pair.
2. Let Rovers rove for 10 minutes, then have partners change and continue 10 minutes more.

Get Connected – Free!

Get the *GEMS Network News*,
our free educational newsletter filled with...

- **updates** on GEMS activities and publications
- **suggestions** from GEMS enthusiasts around the country
- **strategies** to help you and your students succeed
- **information** about workshops and leadership training
- **announcements** of new publications and resources

Be part of a growing national network of people who are committed to activity-based math and science education. Stay connected with the **GEMS** *Network News.* *If you don't already receive the* **Network News,** *simply return the attached postage-paid card.*

For more information about GEMS call (510) 642-7771, or write to us at GEMS, Lawrence Hall of Science, University of California, Berkeley, CA 94720-5200, or gems@uclink.berkeley.edu.

Please visit our web site at www.lhsgems.org.

GEMS activities are effective and easy to use. They engage students in cooperative, hands-on, minds-on math and science explorations, while introducing key principles and concepts.

More than 70 GEMS Teacher's Guides and Handbooks have been developed at the Lawrence Hall of Science — the public science center at the University of California at Berkeley — and tested in thousands of classrooms nationwide. There are many more to come — along with local GEMS Workshops and GEMS Centers and Network Sites springing up across the nation to provide support, training, and resources for you and your colleagues!

Get Connected!
www.lhsgems.org

Yes!

Sign me up for a free subscription to the

GEMS Network News

filled with ideas, information, and strategies that lead to
Great Explorations in Math and Science!

Name_____

Address_____

City_____ State_____ Zip_____

How did you find out about GEMS? (Check all that apply.)
❑ word of mouth ❑ conference ❑ ad ❑ workshop ❑ other: _____
❑ In addition to the *GEMS Network News*, please send me a free catalog of GEMS materials.
❑ Also, sign me up for the online edition of the *GEMS Network News* at this
e-mail address:_____

GEMS
Lawrence Hall of Science
University of California
Berkeley, CA 94720-5200
(510) 642-7771

Ideas ◄
Suggestions ◄
Resources ◄

that lead to Great Explorations
in Math and Science!

Sign up now for a free subscription to the *GEMS* Network News!

101 LAWRENCE HALL OF SCIENCE # 5200

L-61571-25775-62-X

BUSINESS REPLY MAIL
FIRST-CLASS MAIL   PERMIT NO 7   BERKELEY CA

POSTAGE WILL BE PAID BY ADDRESSEE

UNIVERSITY OF CALIFORNIA BERKELEY
GEMS
LAWRENCE HALL OF SCIENCE
PO BOX 16000
BERKELEY CA 94701-9700

Get Connected!
www.lhsgems.org

ASSESSMENT SUGGESTIONS

Anticipated Student Outcomes

Note: Outcomes 1–5 relate to Part I of this GEMS unit. If you also present Part II of the unit, all outcomes (1–8) apply.

1. Students gain and are able to articulate a basic understanding of simple electric circuits by lighting a bulb; demonstrating a closed and open circuit; building a variety of circuits; identifying short circuits, conductors, and insulators.

2. Using schematic diagrams, students are able to represent closed, open, and short circuits, switches, batteries, lightbulbs, and conductors. They are able to build real-life circuits from schematic diagrams and make diagrams of real-life circuits.

3. Students are able to design a simple electric switch.

4. Students are able to grasp the idea of an electric current, and are able to explain that adding batteries to a circuit makes the current stronger.

5. Students gain insight into the idea that an electric circuit is a circular flow of electrons through conductors.

6. Students acquire understanding of the concept of resistance. They are able to explain that anything that slows current down, such as a lightbulb filament, has resistance. They are able to explain that electric current will travel through conductors where it encounters the least resistance.

7. Students are able to demonstrate a more sophisticated understanding of electric circuits, including series and parallel circuits, by assembling and schematically representing examples of these circuits—first without then with switches.

8. Students are able to invent, design, and explain an electric device and represent its circuits using schematic diagrams.

Embedded Assessment Activities

Electrical Inventor's Journal. The Electrical Inventor's Journal is an excellent and holistic assessment instrument over the course of the unit. It includes instruction and recording sheets, writing prompts, and places where students record their ideas and reasoning. Student journal work can provide useful information on their prior knowledge (and alternate conceptions); how well they are grasping the concepts; what questions they have; and how they are preparing for their inventions, so teachers can provide feedback on their plans. The journals include all data sheets, providing the teacher direct access to each student's work throughout the unit. (Addresses outcomes 1–8)

Invented Switches. During Session 5, students are challenged to invent a switch, build it, and represent it with a schematic diagram. In Session 6, students share their switch inventions with one another, as student pairs are divided into alternating roles of "Rover" and "Explainer." This is a great opportunity for the teacher to observe the "Explainers" and assess their comprehension. (Outcomes 1–5)

Electrical Inventions. During Session 10, students are challenged to invent an electric device, build it, and represent it with a schematic diagram. In Session 11, the students share and explain their electrical inventions with one another. This is another excellent opportunity to assess their understanding. (Outcomes 1–8)

Additional Assessment Ideas

Circuit Science: Assessment Activity for the Whole Unit. The following whole-guide assessment activity draws out your students' overall comprehension and skills acquisition, allowing them to apply what they've learned throughout the unit and providing you an opportunity to assess their understanding.

The first part, "What's Wrong with these Circuits?" (page 242), elicits students' understanding of concepts from **Part I** of the unit. Students are given a worksheet with either true-to-life drawings or schematic diagrams of a variety of circuits. Each circuit has something wrong with it. Students explain why each one is flawed, then draw it again to show how the circuit will work. To make this a more complete assessment for your students, include an example of each circuit made with real materials, for students to manipulate, in addition to filling in the worksheet. (Outcomes 1–5)

The second part of this assessment, "Which Circuit?" (page 243), assesses students' understanding of concepts in **Part II.** Students analyze two drawings to decide if each is a parallel or series circuit then explain why. For each drawing, students are asked to draw a schematic diagram of the circuit. Then students answer two questions about the circuits. (Outcomes 5-8)

My Discovery. Students write up and make both true-to-life and schematic drawings to explain a discovery they made during their electrical explorations. (Outcomes 2, 4, 5, 8)

Story of an Electron. Students write the story of an electron in an electric circuit. They are asked to include lightbulbs, batteries, and switches in their stories. (Outcomes 3, 5, 8)

Advertisement Posters. After Session 6, students could design posters advertising their switches, including electrical vocabulary and ideas. After Session 11, students could design posters advertising their electrical inventions, including electrical vocabulary and ideas. (Outcomes 3, 8)

Circuit Science: Part I
What's Wrong with these Circuits?
For each diagram or illustration, tell why the circuit will not work.
Make a new illustration or diagram showing a working circuit.

Circuit	Explain why the circuit doesn't work.	New drawing to show working circuit.
A simple closed circuit.		
A circuit with a switch.		
A circuit with foil "wires."		
A circuit with two batteries.		
A circuit with current flowing through different objects. paper clip coin spoon rubber band		
A circuit with a switch.	What would happen if the switch was closed?	

Circuit Science: Part II
Which Circuit?

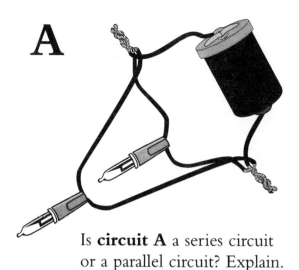

A

Draw a schematic diagram of **circuit A.**

Is **circuit A** a series circuit or a parallel circuit? Explain.

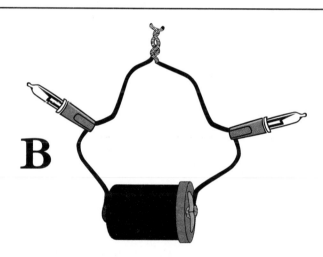

B

Draw a schematic diagram of **circuit B.**

Is **circuit B** a series circuit or a parallel circuit? Explain.

In which circuit will the lights shine brighter? Why?

In which circuit will the battery last longest? Why?

RESOURCES & LITERATURE CONNECTIONS

Sources

Please refer to the "Acquiring and Preparing Materials" section on page 190 for information on the material you'll need for this unit as well as ideas on how to obtain the material.

Related Curriculum Material

Electric Safety

Enterprise for Education
Santa Monica, CA
(888) 300-9864

The "Electric Safety" series is one of several related to electricity published by Enterprise for Education. Primarily for use by utility companies, these publications help K–12 students learn about the sources and uses of energy, electric safety, energy efficiency, and other topics.

The five titles in the "Electric Safety" series are: *Leo Learns Electric Safety*, an interactive video game—adapted for a variety of grade ranges—to teach electric safety; *Mouse House Surprise,* a story book which introduces students to the utility's role in providing reliable electric power, what it's used for, and how to use it safely; *Sam and Jody Save the Circus,* another story book which provides students with opportunities to apply what they know about electricity and its dangers, as well as to learn more electric safety "do's" and "don'ts;" *Paths for Electricity,* a workbook plus copper wire and flashlight bulbs with which students can investigate circuits, conductivity, insulators, short circuits, and grounding, as well as predicting the path electricity will take; and *An Electrical Engineer Plans a Safe Outdoor Concert,* a book through which students work in teams and apply many math skills to determine how much electricity

and how many generators are needed for a rock concert—thus learning about branch circuits, overloading, and the role of circuit breakers and fuses. All publications come with teacher's guides, and together the materials develop basic reading, writing, math, and computer skills, using electric safety as the content. Learn more about this electric safety series and many others at www.entfored.com.

Multicultural Women of Science
Three Centuries of Contributions with Hands-On Experiments and Activities for 37 Weeks

by Leonard Bernstein, Alan Winkler, and Linda Zierdt-Warshaw
The Peoples Publishing Group, Inc., Saddlebrook, NJ
(1996; 178 pp.)

Designed to provide one lesson per week for the entire school year, this book documents the major impact minority women have had on many scientific fields. Each lesson contains a biographical sketch, a hands-on activity, and a "Think Work Act" page with critical thinking questions and at least four activities featuring a variety of skills and learning styles. The stories show how these women overcame social and economic obstacles, thus presenting positive role models. The back matter contains a glossary, more biographical sketches, suggested readings, and an index. An accompanying teacher's guide is also available.

Nonfiction for Students

Accidents May Happen
by Charlotte Foltz Jones
Delacorte Press, New York, NY
(1998; 96 pp.)

From the author of *Mistakes That Worked,* this book collects a range of stories demonstrating how scientific discoveries and inventions often happen by accident. Readers learn the origins of a variety of products we take for granted, including the yo-yo, cellophane, liquid paper, and Avon cosmetics.

Always Inventing
A Photobiography of Alexander Graham Bell
by Tom L. Matthews and Gilbert Grosvenor
National Geographic Press, Washington, D.C.
(1999; 64 pp.)

Filled with lots of black-and-white photographs, this book traces the life and many inventions of one of the 20th century's most creative geniuses.

Brainstorm!
The Stories of Twenty American Kid Inventors
by Tom Tucker
Farrar, Straus and Giroux, New York, NY
(1998; 144 pp.)

Recounts the stories of a number of inventions created by children. Ranging from a safety device for power looms to the Popsicle—which was invented by an 11-year-old boy—the book features a wide range of inventors, both male and female.

Electricity
by Steve Parker
Dorling Kindersley, New York, NY
(2000; 64 pp.)

After two centuries of research and experimentation, electricity has revolutionized the way we live and become an indispensable part of scientific progress. This book brings vividly to life our exploration of this powerful natural force. It is a vital part of the living world. Our eyes receive light rays and turn them into tiny electrical signals that pass along our nerves to the brain. The chemicals in batteries create power for such things as toys and flashlights. The invention of electric motors enabled the creation of a whole range of household appliances that have shaped the modern world. See how electricity works nerves and muscles and what causes lightning. Learn how electricity can transmit voices and images across an ocean and what makes a silicon chip work. Discover the amazing speed of charged atoms, what happens when you tune in the radio, and much, much more!

Girls Think of Everything
Stories of Ingenious Ideas by Women
by Catherine Thimmesh
Houghton Mifflin, New York, NY
(2000; 64 pp.)

A collection of inspirational portraits of women and girls who changed the world with their inventions, this book surveys unique and creative products of ingenuity and hard work.

How Ben Franklin Stole the Lightning
by Rosalyn Schanzer
HarperCollins, New York, NY
(2003; 33 pp.)

Focuses on Benjamin Franklin's role as an inventor of whimsical gadgets and practical contraptions, with an emphasis on his experiment of flying a kite during a rainstorm.

Inventing the Future
A Photobiography of Thomas Alva Edison
by Marfé Ferguson Delano
National Geographic Press, Washington, D.C.
(2002; 64 pp.)

Filled with fascinating photographs, this book traces Edison's life and career. Throughout, the author emphasizes Edison's curiosity and the joy he derived from discovery.

Inventors
by Martin W. Sandler
HarperCollins, New York, NY
(1996; 93 pp.)

Americans have been characterized by their inventive spirit since the days of Benjamin Franklin, but the nineteenth and early twentieth centuries proved especially fruitful in groundbreaking discoveries that revolutionized life as we know it. This richly illustrated Library of Congress book presents the evolution of these inventions as it has never been seen before—and celebrates the spirit of the great American inventors who let loose their imaginations and changed the world forever.

The Kid Who Invented the Popsicle
And Other Surprising Stories about Inventions
by Don L. Wulffson
Puffin, New York, NY
(1999; 128 pp.)

This book offers brief descriptions of how 114 different, everyday objects came to be. Each interesting invention—from animal crackers to miniature golf to zippers—is covered in a short, one-page tidbit.

The Kid Who Invented the Trampoline
More Surprising Stories About Inventions
by Don L. Wulffson
Dutton Books, New York, NY
(2001; 120 pp.)

This sequel to *The Kid Who Invented the Popsicle* includes the stories behind how leotards, Post-it notes, parking meters, and 47 other everyday objects were invented.

The Magic School Bus and the Electric Field Trip
by Joanna Cole;
illustrated by Bruce Degen
Scholastic, New York, NY
(1997; 48 pp.)

Ms. Frizzle and her class explore electricity by taking a trip through the town's electrical wires. Subjects include how electricity is made in power plants, how transformers function, and what constitutes differences in voltage.

Mistakes That Worked
40 Familiar Inventions and How They Came To Be
by Charlotte Foltz Jones;
illustrated by John O'Brien
Doubleday, New York, NY
(1994; 96 pp.)

From the author of *Accidents May Happen,* this book details the haphazard beginnings of a number of every-

day products, including potato chips, Scotchguard, Silly Putty, and X-rays.

The New Way Things Work
by David MacAulay
Houghton Mifflin, New York, NY
(1998; 400 pp.)

This hefty volume has a chapter on electricity that addresses subjects such as circuits, switches, and batteries. It also has separate sections on a wide range of electric devices, like the electric light, electric guitar, and electric motor.

Switch On, Switch Off
by Melvin Berger;
illustrated by Carolyn Croll
Harper Trophy, New York, NY
(2001; 32 pp.)

Explains how electricity is produced and transmitted, how to create electricity using an electrical wire and a magnet, how generators supply electricity for cities, and how electricity works in homes. A good introduction for younger students.

The Way Science Works
by Robin Kerrod and Sharon Holgate
Dorling Kindersley, New York, NY
(2002; 160 pp.)

Find out why science matters and how it is changing our world. Contains the latest examples of science in action together with fun facts and figures, practical experiments, and a full glossary and index. Test theories in more than 100 hands-on projects. Next-generation visuals and cutting-edge content help illuminate key scientific developments. An exciting way to keep ahead of the curriculum and discover science for yourself.

Women Invent!
Two Centuries of Discoveries That Have Shaped Our World
by Susan Casey
Chicago Review Press, Chicago, IL
(1997; 152 pp.)

Offers portraits of a wide range of successful women and girl inventors, and takes the reader on a step-by-step journey through the process of inventing—from coming up with an idea and creating a model to obtaining a patent and having it manufactured and sold. Included are the stories of many school-aged children who have been granted patents. Resource guide provides information on Internet sites, inventors camps, programs, and contests.

Fiction for Students

Ben and Me
A New and Astonishing Life of Benjamin Franklin as Written by His Good Mouse Amos
by Robert Lawson
Little, Brown and Company, Boston, MA
(1988; 114 pp.)
Grades 3-5

Benjamin Franklin's companion, Amos the mouse, recounts how he was responsible for Franklin's inventions and discoveries such as the Franklin stove, his experiments with electricity, and the lightning rod. Franklin's adventures with "Poor Richard's Almanac" and the Revolutionary War are retold in a humorous fashion.

Einstein Anderson Sees Through the Invisible Man
by Seymour Simon;
illustrated by Steven D. Schindler
William Morrow, New York, NY
(1998; 96 pp.)
Grades 4-7

In Chapter 7, "A Cold Light," Einstein's ever hopeful friend Stanley is excited about a secret formula for cold light, which would offer an alternative to electric light. Einstein explains that light energy can be chemically produced by using a known formula similar to the substance that causes a firefly's luminescence, but production would be too expensive.

Einstein Anderson Shocks His Friends
by Seymour Simon;
illustrated by Fred Winkowski
Viking Press, New York, NY
(1980; 73 pp.)
Grades 4-7

In the first chapter, "The Electric Spark," Einstein uses static electricity to scare off his nemesis Pat Burns.

Thunder Cake
by Patricia Polacco
Philomel Books, New York, NY
(1997; 32 pp.)
Grades Preschool-6

Grandma finds a way to dispel her grandchild's fear of thunderstorms. The two race to gather ingredients and bake a "thunder cake" before the storm arrives. By counting the seconds between when they see the lightning and hear the thunder, they are able to measure the distance of the approaching storm. Both counting and cake-making provide an engaging and empowering distraction from the energy, sounds, and sights of the thunderstorm. The recipe for thunder cake is included.

The Toothpaste Millionaire
by Jean Merrill;
illustrated by Jan Palmer
Houghton Mifflin, Boston, MA
(1999; 96 pp.)
Grades 2-8

Incensed at the price of toothpaste, Rufus sets out to make and then sell his own inexpensive brand. While the book does not dwell on the subject of invention, it does address how to transform an invention into a business. Throughout the book are many opportunities to use math—nicely illustrating the need for and use of mathematics in real-world problem solving.

Videos

Inventions That Changed Our Lives
The Wonder of Electricity
Goldhil Home Media
(2000; 27 minutes)

Inventions That Changed Our Lives, a six-volume series, uses modern-day technological inventions to take a fascinating peek back at early inventors and their creations. Each episode focuses on a different area of inventions—communication and transportation, ancient inventions, medicine, plastics, warfare, and electricity. *The Wonder of Electricity,* chronicles a century of timesaving and entertaining devices such as the washing machine, refrigerator, camera, radio, television, and computer.

Understanding Electricity
Discovery Channel School
(1997; 51 minutes)

From the individuals who use it to the engineers who harness it, this video examines the power and purposes of electricity. It also covers how electricity is generated and delivered to homes, as well as how ideas about this force may combine the electricity in our bodies with man-made electrical systems.

CD-ROMs

Electricity
Discovery Channel School

Students learn the basics of electricity from classroom demonstrations and experiments. This interactive CD-ROM makes static, current, and circuit electricity come alive. Build a battery with pennies, learn about Ben Franklin's experiments, or build a telegraph. Includes a teacher's guide with over 100 pages of teacher support material.

Exploring Physics
Electricity and Magnetism
Chandra Publications, LLC

Developed with support from the National Science Foundation, this CD has hands-on activities for middle grades (5–9). Six topics (Batteries, Bulbs and Switches; Understanding Batteries; Simple Circuits; Static Electricity; Resistors and Capacitors; and Magnets and Magnetism) are supported by concept activities, reading pages, and progressively more challenging activities. A vast repertoire of links provides easily accessible background information—with attractive pictures, animations, and examples. For more information, visit www.exploringphysics.com.

Internet Sites

While we do our best to provide long-lived addresses in this section, websites can be mercurial! Comparable alternative sites can generally be found with your Web browser.

About Inventors
http://inventors.about.com/

Begin here to connect to many sites (such as 20th Century Inventions, Find: A to Z Inventions/Inventors, Famous Inventions/Inventors, and Technology Timelines) each with hundreds of links.

The Atoms Family
www.miamisci.org/af/sln/

This resource contains educational activities relating to different forms of energy. The website is based on The Atoms Family exhibit at The Miami Museum of Science.

The Basics of Electric Circuits
www.windows.ucar.edu/spaceweather /how_circuit_works.html

Provides a mechanical analogy for understanding electric current.

Electrical Safety World
www.culverco.com/resources/esw/index.html

Designed by a communications company to help utilities educate their customers, this site is full of great information about electricity—what it is, how it's generated, how it travels from power plant to homes and into appliances, and how to use it safely. Pop-up windows define key terms, and games and experiments are included.

Physics 2000
www.colorado.edu/physics/2000/index.pl

Visitors to this site take an interactive journey through modern physics, and learn visually and conceptually about 20th century science and high-tech devices such as X-rays, microwave ovens, lasers, and laptops. Contains a substantial amount of advanced content, but it's carefully explained in a highly user friendly format. The "Electromagnetic Waves and Particles" portion of the Science Trek section contains good information and activities for students ready to learn more about the physics of electricity.

REVIEWERS

We warmly thank the following educators, who reviewed, tested, or coordinated the trial tests for *Algebraic Reasoning, Electric Circuits, Invisible Universe,* and *Living with a Star* in manuscript or draft form. Their critical comments and recommendations, based on classroom presentation of these activities nationwide, contributed significantly to this GEMS publication. (The participation of these educators in the review process does not necessarily imply endorsement of the GEMS program or responsibility for statements or views expressed.) Classroom testing is a recognized and invaluable hallmark of GEMS curriculum development; feedback is carefully recorded and integrated as appropriate into the publications. WE THANK THEM ALL! ■

ARIZONA

Liberty School District #25, Buckeye
Wayne Bryan★
Terri Matteson

Arrowhead Elementary School, Phoenix
Noel Fasano
Jorjanne Miller
Kimberly Rimbey★
Delores Salisz
George Sowby
Coreen Weber

Hohokam Middle School, Tucson
Maria Federico-Brummer
Jennine Grogan★

Pistor Middle School, Tucson
Mike Ellis★

ARKANSAS

Carl Stuart Middle School, Conway
Chris Bing
Linda Dow
Gene Hodges
Charlcie Strange★

CALIFORNIA

Albany Middle School, Albany
Kay Sorg★

Rio Del Mar Elementary School, Aptos
Chris Ferrero
Doug Kyle
Tom LaHue★
Debbie Lawheed

Endeavor Elementary, Bakersfield
Matthew Diggle
Jan Karnowski★
Carolyn Reinen
Julie Rosales

Le Conte Elementary, Berkeley
Carole Chin★
Lorna Cross
Jennifer Smallwood

Longfellow Middle School, Berkeley
Karen Bush★

Juan Crespi Junior High School, El Sobrante
Randa Emera★
Juli Goldwyn
Geri Lommen
Julie Skow★

Oak Manor School, Fairfax
Celia Cuomo

Lorin Eden Elementary School, Hayward
Donna Goldenstein★
Elise Tran

M.H. Stanley Intermediate School, Lafayette
Tina Woodworth

Altamont Creek Elementary, Livermore
Pauline Huben★
Janice Louthan

Emma C. Smith Elementary, Livermore
J. Gulbransen

Leo R. Croche Elementary, Livermore
Corinne Agurkis

Mammoth Elementary School, Mammoth Lakes
Sue Barker★
Sandy Bramble
Stacey Posey
Janis Richardson

Cypress Elementary, Newbury Park
Cheryl Bowen
Christina Myren★
Kim Thompson

Bret Harte Middle School, Oakland
Anthony Cody

Roosevelt Middle School, Oakland
Ileana de la Torre
Suzanne Frechette
Elizabeth (Betsy) Rosenberg★

Sobrante Park Elementary, Oakland
Teri Hudson★
Paul McDermott
Raul Nunez
Julie Pokrivnak

Westlake Middle School, Oakland
Hindatu Mohammed★

Los Medanos Elementary, Pittsburg
Tanya Duke★
Syglenda Ford
Karen Staats
Delores Williams

Creekside Middle School, Rohnert Park
Emily Dunnagan
Al Janulaw★
Bonny Stene

Baldwin Elementary, San Jose
Beth Harris
Marietta Harris
Brenda Hough★
Shelley McCracken

Downer Elementary School, San Pablo
Antoineta Franco
Melinda Melaugh
Diana Ortega
Amy Scott
Emily Vogler

Rincon Valley Middle School, Santa Rosa
Jim Bennett
Penny Sirota★
Laurel VarnBuhler

Acacia Elementary, Thousand Oaks
Banny Anderson
Karen Barker
Cathy Bostic
Sharon Sickler

Meadows Elementary, Thousand Oaks
Laura Nedwick

Glen Cove Elementary, Vallejo
Cindy Jones
Diana McKeever
Charles Shannon★

COLORADO

Cory Elementary School, Denver
Debbie Beard
Scott Sala★

FLORIDA

Howard Middle School, Orlando
Hassan Champion
Susan Leeds★
Nicole Ryker

ILLINOIS

Sandwich Middle School, Sandwich
Angela Knierim★
Krista Olson
Kim Paulus
Amy Sigler

OHIO

Cameron Park Elementary, Cincinnati
Cris Cornelssen
Kris Thompson

Lakeside Elementary, Cincinnati
Bob Flinn★
Stacey Owens

Jewish Education Center of Cleveland, Cleveland Heights
Elaine Feigenbaum
Ida Friedman-Kasdan
Eugenia Johnson-Whitt★
Jill Leve
Rabbi Reich

Fairfield North Elementary School, Hamilton
Carol Gregory
Carolyn Kolkmeyer
Sheila Messersmith
Wendy New★
Sheila Webb

Wilmington Middle School, Wilmington
Jeff Bourne
Gary Downing
Sue Hanna
Amy Steinle★
Kathy Vincent

OREGON

Redwood Elementary School, Grants Pass
Shelly Brandes
Lorelei Dean
Renee Grant★
Marleen Knight

Kennedy Elementary, Medford
Carol Fitspatrick
Teena Staller★

Lone Pine Elementary, Medford
Madolyn Malloy

Oak Grove Elementary, Medford
Cheryl Lemke

Portland Lutheran School, Portland
Karin Maier★
Ruth Mannion
Desi Pritchard
Tom Zuch

PENNSYLVANIA

Crossroads Middle School, Lewisberry
Carol Brame
Hannah Leigey★

Fairview Elementary School, New Cumberland
Beth Erikson
Julie Estep
Louise Shuey
Sue White

SOUTH CAROLINA

McCormick Elementary School, Mullins
Barbara Baker
Debbie Beeson
Sarah Dew
Patricia Grant
Brenda Ladson
Fannie Mason★

College Park Middle School, Summerville
Dottie Adams
Marsha Lindsay★

Sangaree Intermediate School, Summerville
Deanna Hefner
Sandy Wiedmeyer★

TENNESSEE

Bartlett Elementary School, Bartlett
Penny Blair
Jenny Carter
Joyce Cornett
Jill Crumpton★
Anne Faulks★
LaJuana Heaston
Lee Loft
Portia Tate
Jenny Underwood

TEXAS

Bammel Elementary, Houston
Anna Dugger
Tracey Harros★
Staci Horan
Cindy Lane

Clear Creek ISD, Seabrook
Christine Casaburri
Susan Lallo
Ann Martinez
Katherine Mays
Sandy Peck★
Sally Wall

WASHINGTON

Covington Middle School, Vancouver
Laurie Cripe★
Bryce Hampton
Marie Morasch
Jackie Serniotti

★Trial Test Coordinator